Eye-Tracking with Python and Pylink

Zhiguo Wang

Eye-Tracking with Python and Pylink

Springer

Zhiguo Wang
Zhejiang University
Hangzhou, Zhejiang, China

ISBN 978-3-030-82634-5 ISBN 978-3-030-82635-2 (eBook)
https://doi.org/10.1007/978-3-030-82635-2

This Springer imprint is published by the registered company Springer Nature Switzerland AG
The registered company address is: Gewerbestrasse 11, 6330 Cham, Switzerland

Preface

Python is becoming more and more popular among scientists. Its application in psychology has expanded from scripting computer-based experimental tasks to all aspects of research, including but not limited to data collection, analysis, and visualization. I had little knowledge of computer programming when I started my graduate study about 17 years ago. I experimented with different programming languages and tools but finally settled down with Python for its easy syntax, flexibility, versatility, and open community.

The content of this book is based mainly on my research experience and the courses I have taught previously. There are quite a few Python books on the market featuring programming tools designed for psychologists (e.g., PsychoPy). What sets this book apart is its focus on eye-tracking.

Eye-tracking is a widely used research technique in psychology and neuroscience labs. The eye-trackers used in the lab are typically faster, more accurate, and of course, more expensive than the ones seen in consumer goods (e.g., VR goggles) or usability labs. The eye-trackers featured in this book are the EyeLink series trackers manufactured by SR Research Ltd (Canada). The EyeLink eye-trackers are arguably the best research-grade tracker available on the market. This high-speed eye-tracker has unmatched precision and accuracy, and it has been used in well over 9000 peer-reviewed journal articles.

This book will first introduce the building blocks and syntax of Python, and then discuss libraries that we can use to program psychology experiments, that is, PsychoPy and Pygame. For eye-tracking, this book will cover the Pylink library in detail. The example scripts accompanying this book are freely available on GitHub, https://github.com/zhiguo-eyelab/Pylink_book. This book is a useful reference for eye-tracking researchers, but you can also use it in graduate or undergraduate level programming courses.

Special thanks go to Dr. Sam Hutton, Dr. Jiye Shen, Dr. Jon Peirce, and Dr. Sebastiaan Mathôt for their valuable feedback on an earlier version of this book.

Hangzhou, China Zhiguo Wang

Preface

Contents

Chapter 1
A Gentle Introduction to Python

Contents

This chapter will focus on the basic features of the Python programming language necessary for implementing simple psychological experiments. For a more comprehensive introduction to Python, I recommend the tutorial on the official Python website (https://docs.python.org), especially if you have little programming experience. The free online book *Dive into Python* by Mark Pilgrim is also a very helpful reference.

Z. Wang, *Eye-Tracking with Python and Pylink*,
https://doi.org/10.1007/978-3-030-82635-2_1

Installing Python

There are many different Python "distributions" (e.g., Anaconda, Canopy), which bundle features and tools that are not part of the official Python distribution. This book will feature the official Python distribution that is freely available from https://www.python.org. The example scripts presented in this book are based on Python 3 (also available from https://github.com/zhiguo-eyelab/Pylink_book).

Windows

To install Python 3 on a Windows PC, please download the installer, run it, and tick the "Add Python 3.x to PATH" and "Install launcher for all users" options. The Python launcher is a convenient tool when you have multiple versions of Python on your Windows PC (Fig. 1.1).

macOS and Linux

To verify if you have Python 3 on your Mac, launch a terminal and type *python3* at the command-line prompt. As shown in the shell output below, my MacBook has Python 3.6.6. If you do not have Python 3 on your Mac, please download the Mac OS X installer from https://www.python.org and install it.

```
Zhiguos-MacBook-Pro:~ zhiguo$ python3
Python 3.6.6 (v3.6.6:4cf1f54eb7, Jun 26 2018, 19:50:54)
[GCC 4.2.1 Compatible Apple LLVM 6.0 (clang-600.0.57)] on darwin
Type "help", "copyright", "credits" or "license" for more information.
>>>
```

If you use Ubuntu or another Debian-based Linux distribution, please run the following command in a terminal to install Python 3.6 or a later version.

```
sudo apt install python3.6
```

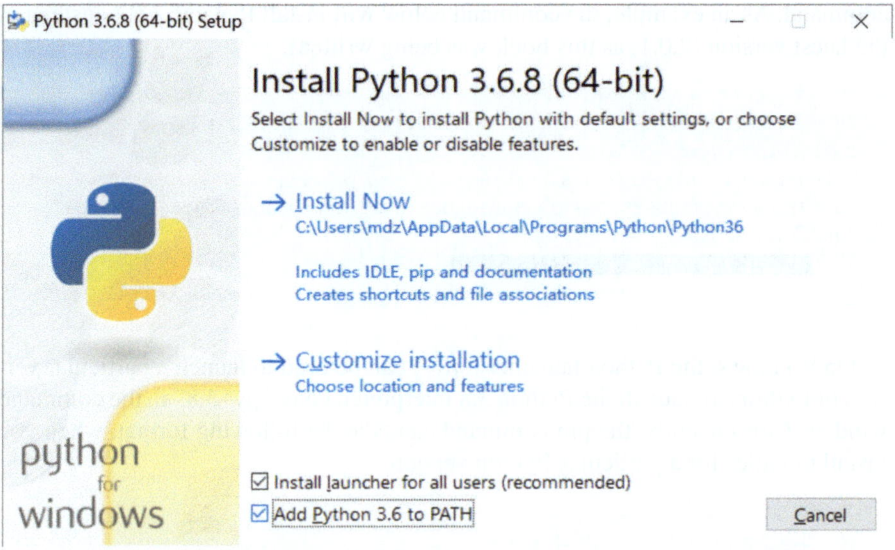

Fig. 1.1 Python installation options on Windows

Installing Python Modules

Thousands of Python modules are actively developed and maintained by the community. These modules have greatly expanded the functionality of Python and allow users to access solutions others have already created for common problems. The preferred installer for Python modules is *pip*, which is now a module included in the official Python distribution. With *pip*, users can search, download, and install Python modules from either a public (e.g., Python Package Index, or PyPI) or a private online repository. On macOS and Linux systems, the pip command takes the following format. Here, the "-m" option requests the Python interpreter to look for the module that follows "-m" (i.e., "pip" in the command below) and execute it.

```
python -m pip install SomePackage
```

The above command will retrieve and install a module for the default Python on your computer. To install modules for a particular version of Python, include the version number in the "python" command.

```
python3.x -m pip install SomePackage
```

The pip command, by default, will install the latest version of a module. If a particular version of a module is needed, include the version number in the command. As an example, the command below will install Pygame 1.9.6 instead of the latest version (2.0.1, as this book was being written).

```
Zhiguos-MacBook-Pro-2:~ zhiguo$ python3.7 -m pip install pygame==1.9.6
Collecting pygame==1.9.6
  Downloading https://files.pythonhosted.org/packages/32/37/453bbb62f90feff
2a2b75fc739b674319f5f6a8789d5d21c6d2d7d42face/pygame-1.9.6-cp37-cp37m-
macosx_10_11_intel.whl (4.9MB)
    |████████████████████████████████| 4.9MB 6.8MB/s
```

On Windows, the Python launcher ("py") can be used to launch a particular version of Python. To launch the Python 3.8 interpreter, enter "py -3.8" in the command window. Consequently, the pip command can take the following format, when you install modules for a particular Python version.

```
py -3.x -m pip install SomePackage
```

Python Shell

Python is an interpreted language, which means users do not need to deal with source code linking and compiling, as in languages like C/C++. In other words, you do not need to compile the source code to generate executable files, but rather the source code is evaluated line by line by an interpreter, which converts the source code into instructions recognizable by the computer hardware. In the simplest case, you send a statement (command) to the Python interpreter; the Python interpreter evaluates the statement and sends back its response. When used interactively in a Python *shell*,[1] the Python interpreter allows you to experiment with the various features of the language or test functions during program development.

On macOS and Linux, entering *python3* at a command-line prompt will evoke the Python interpreter. The interpreter first prints a welcome message, starting with its version number, and then the prompt (">>>"—three angle brackets), which indicates that the Python interpreter is ready to accept commands.

```
Zhiguos-MacBook-Pro:~ zhiguo$ python3
Python 3.6.6 (v3.6.6:4cf1f54eb7, Jun 26 2018, 19:50:54)
[GCC 4.2.1 Compatible Apple LLVM 6.0 (clang-600.0.57)] on darwin
Type "help", "copyright", "credits" or "license" for more information.
>>>
```

[1]A shell is a user interface for accessing the computer operating system services. The use of a shell usually involves entering a single command and then receiving the relevant response from the operating system (typically printed in the shell as messages). The Python shell works in the same way, but processes Python commands instead of operating system commands.

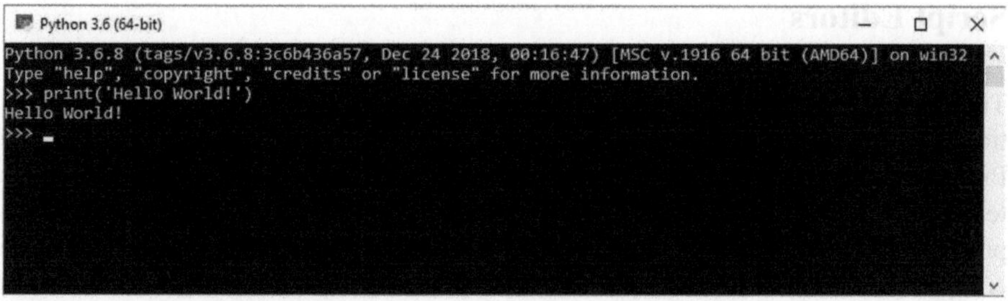

Fig. 1.2 A Python shell is included in the Windows version of the official Python distribution; it is accessible from the Start menu

To quit the Python shell, enter quit() or exit(), or simply press Ctrl-D.

```
>>> quit()
```

On Windows, a Python shell (see Fig. 1.2) is accessible from the Start menu. You can also use a Windows command prompt to start a Python shell.

You can use Python in an interactive manner in a Python shell. If you enter a command or a statement, the Python shell will print the return value.

```
>>> 3 + 4
7
```

Typing *help* in the Python shell will show a notice that you can access the help information with either the *help()* function or the interactive help utility. In the Python shell below, we retrieve the help information for the "print" function with *help(print)*.

```
>>> help
Type help() for interactive help, or help(object) for help about object.
>>>
>>> help(print)
Help on built-in function print in module builtins:
print(...)
    print(value, ..., sep=' ', end='\n', file=sys.stdout, flush=False)
```

Script Editors

The Python shell helps test out basic features and functions, but real-world applications usually involve many lines of code. It is impractical to type code into the Python shell line by line, so the code is typically placed into a Python script, i.e., a text file that ends with a ".py" extension. By sending a script to the Python interpreter, far more complex tasks than simple calculations can be performed, such as a computer-based reaction time task.

You can use any text editor (e.g., Vim, Emacs, NotePad++, etc.) to write your Python script. There are also feature-rich development tools that offer a script editor and additional features like code debugging (e.g., PyCharm, Spyder, Visual Studio Code).[2] These tools are known as integrated development environments, or IDEs for short, and are indispensable to large-scale software development. For the relatively short scripts required to implement a typical psychology experiment, a full-fledged IDE is usually unnecessary. Which editor or IDE to use is very much a personal

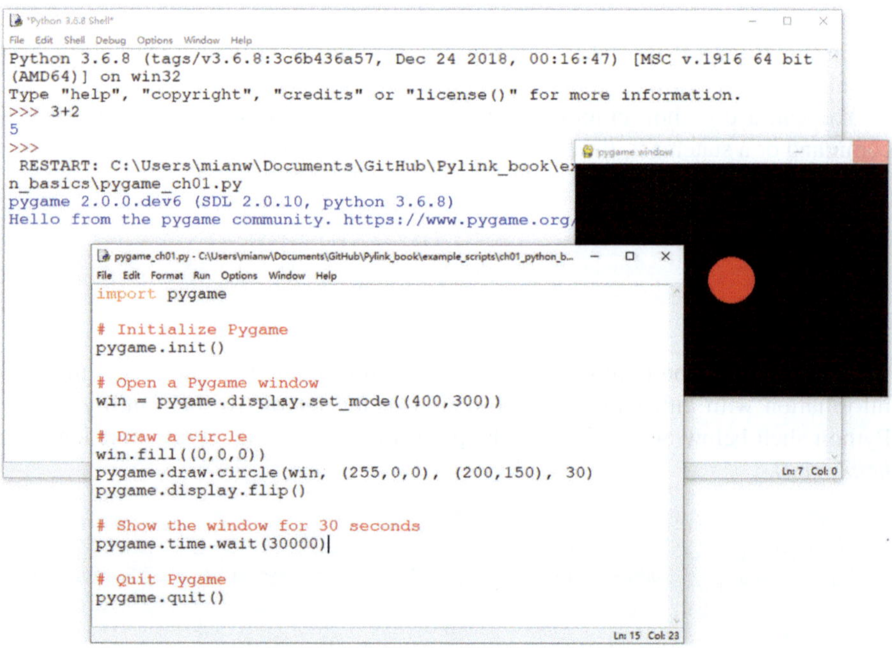

Fig. 1.3 IDLE: a simple Python IDE with a shell and a script editor. The script shown here will open a window and then draw a red disk

[2] See the Python Wiki for a list of Python IDEs, https://wiki.python.org/moin/IntegratedDevelopmentEnvironments.

preference. Still, it is worth using a text editor that is Python-aware, i.e., it should have functions like syntax highlighting and autocompletion.[3]

The official Python distribution comes with a basic Python IDE called IDLE. It features both a Python shell and a script editor. By default, launching IDLE will bring up the Python shell; from File -> New, you can then open the script editor. Any lines of code that are typed into the editor need to be saved before they can be passed to the Python interpreter. The saved file should have a ".py" extension. Once the file has been saved, clicking Run-> Run module (or press F5) in the script editor will run the script from beginning to end.

In the short script shown in Fig. 1.3, we first import *pygame* and then use its *display*, *draw*, and *time* modules to present a red disk in a window. There is no need to delve into the details of this script at this point, but it is worth noting that texts following the "#" (hash) character are comments that help improve code readability.

Data Types

Programming, in essence, is data manipulation. In Python, frequently used data types include numbers, strings, lists, tuples, dictionaries, and Booleans.

Numbers

Numbers can be integers (e.g., 2, 4, 20), floats (e.g., 5.0, 1.6), fractions, or complex numbers. The mathematical operations which can be performed on numbers include + (addition), – (subtraction), * (multiplication), / (division), etc. Note that, in Python 3, division always returns a float number, whereas in Python 2, dividing an integer by another integer will return the result of a floor division (e.g., 5/2=2, −5/2=−3), e.g., another integer. In Python 3, floor division is done with the operator "//" only.

Strings

Besides numbers, Python can also manipulate strings or text. Strings can be expressed in single quotes ('...') or double quotes ("...").

[3] Autocompletion can greatly improve coding efficiency. For instance, typing in "math.lo" and then pressing the Tab key in the IDLE shell will list all math functions with names starting with "lo," e.g., log, log2, and log10.

```
>>> 'this is a string'
'this is a string'
>>> "this is a string"
'this is a string'
>>> print("this is a string")
this is a string
>>> print("\" can be used in a string")
" can be used in a string
```

The *print*() function helps produce a more readable output in the shell by omitting the enclosing quotes. Note the last shell command in the above box printed out a double quote at the start of the string. Because the Python syntax dictates that double quotes mark the start or end of a string, we need to let the Python interpreter know that this double quote is just an ordinary double quote in a string. This solution is to escape the double quote with the Python escape character "\" (backslash). This operation is known as "escaping," and it is also seen in many (if not all) programming languages to deal with special characters (e.g., "\n" represents a new line) and characters reserved for syntax purposes. By adding the escape character, "\"" escapes a double quote, and "\\" escapes a backslash. Strings can also contain special formatting characters that also require the escape character. The tab ("\t") and new-line ("\n") characters are good examples.

```
>>> print("\"")
"
>>> print("\\")
\
>>> print("this is a tab\t see it?\nstart a new line")
this is a tab          see it?
start a new line
```

One feature of Python is that strings can be concatenated with the "+" operator and repeated with the "*" operator. In the shell example below, the string "Python" is repeated three times and then gets concatenated or "glued together" with another string "123."

```
>>> 'Python' * 3 + '123'
'PythonPythonPython123'
```

The characters in a string are indexed, with the first character indexed by 0, the second character indexed by 1, etc. The indices are negative when going in the right-to-left direction. We can use a single index (in a pair of square brackets) to retrieve a character, or an index range to extract multiple characters.

```
>>> s = "Python"
>>> s[0]
'P'
>>> s[-2]
'o'
>>> s[0:3]
'Pyt'
```

In the last shell command shown above, the index range [0:3] retrieves the first three characters from the string "Python"—"Pyt." You may be wondering why this command does not extract the fourth character, even though its index (3) is in the brackets. It is best to think of an index as an invisible marker representing the element to its right. When going in the right-to-left direction, there is no character right to index 0, so the first character we can retrieve from the string "Python" has an index of −1.

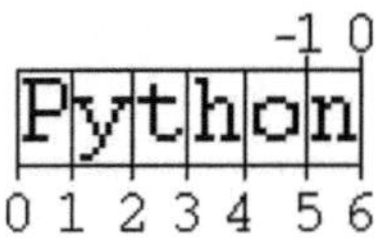

Booleans

Booleans are either *True* or *False*. They are the return values from comparisons and Boolean operations. For instance, the statement "3 < 5" will return *True*; the statement "True and False" will return False, while "True and True" will return *True*. Booleans are frequently used in control flow statements, for instance, in an *if* statement (see the following sections for details).

Booleans can be interpreted as numbers, with *True* = 1 and *False* = 0. So, "True + True" will return 2. You can also perform Boolean operations on other Python data types. For numbers, 0 is *False,* and all nonzero numbers are *True*; with structured data like lists, tuples, sets, and dictionaries (see below), empty ones are *False,* and non-empty ones are *True*. Python also has a constant *None* for a null value; comparing *None* to any other data type will always return *False*.

List

A list is a sequence of comma-separated values of the same or different data types. Items in a list can be accessed through indexing, just like strings.

```
>>> a = [1, 2, 3, 4]
>>> b = ['a', 1, [1, 2], 3.0]
>>> b[2]
[1, 2]
>>> b[-1]
3.0
>>> a + b
[1, 2, 3, 4, 'a', 1, [1, 2], 3.0]
```

As with strings, lists can be concatenated with the "+" operator (see the last shell command above) and repeated with the "*" operator. Lists are mutable. You can change a list, for example, by adding items to the end with *append*() or sorting the items with *sort*(). Other common operations that can be performed on a list include *count*(), *extend*(), *insert*(), *pop*(), *remove*(), etc.

Tuple

A tuple consists of values separated by commas. As with lists, tuples can contain a heterogeneous sequence of elements, e.g., numbers, strings, lists, and tuples. The items in a tuple are usually enclosed in a pair of parentheses (curved brackets), but we can omit the parentheses when passing a tuple to a variable. Unlike lists, tuples are immutable. You cannot remove or replace the elements of a tuple.

```
>>> t = 1, 'a', 3.567, [1,2,3]
>>> t
(1, 'a', 3.567, [1, 2, 3])
```

An empty tuple can be declared with a pair of parentheses (); a single-item tuple can be declared with the item, followed by a comma.

```
>>> t = 0.4,
>>> t
(0.4,)
```

Note the above single-item tuple has a comma in the parenthesis. Without the comma, the Python interpreter will return the value enclosed in the parenthesis.

```
>>> t = (0.4)
>>> t
0.4
```

Tuples can be used to store data that are not supposed to change while the script is running, for instance, the resolution of a monitor.

Dictionary

Dictionaries are sometimes referred to as "associative memories." A dictionary is a set of key-value pairs, with the key and value separated by a colon. An empty dictionary can be instantiated with a pair of curly brackets. The keys can be used to retrieve the associated value from the dictionary. For instance, in the shell command below, *d['A']* returns the value associated with key 'A' in the dictionary *d*, which is 65. To add a new key-value pair to an existing dictionary, simply specify the new key in a pair of square brackets, and pass a value to it, for instance, *d['D'] = 68*.

```
>>> d = {'A': 65, 'B': 66, 'C': 67}
>>> d['A']
65
>>> d['D'] = 68
>>> d
{'A': 65, 'B': 66, 'C': 67, 'D': 68}
```

Somewhat confusingly, you can use the *list*() function to extract the keys of a dictionary. This function returns the keys in the order that the *key-value* pairs were added to the dictionary. To check whether a single key is in the dictionary, use the *in* operator.

```
>>> list(d)
['A', 'B', 'C', 'D']
>>> 'C' in d
True
```

Set

A set contains unique (non-duplicating) and unsorted values; it conforms to the mathematical definition of a set. Sets are defined with curly braces or the *set*() function. Note that to create an empty set, you have to use the *set()* function, not {}; the latter will create an empty dictionary instead.

```
>>> {'a', 'a', 'a'}
set(['a'])
>>> set(['a', 'a', 'a'])
set(['a'])
```

Unlike lists, items in a set must all be unique. Taking advantage of this feature, one can quickly remove duplicates in a list by converting the list into a set and back into a list.

```
>>> x = [1, 2, 3, 4, 1, 2, 4, 5]
>>> list(set(x))
[1, 2, 3, 4, 5]
```

There are a few other handy methods that can be used with sets. For instance, *pop*() will randomly remove an item from a set and return it (selection without replacement); the *union*() method combines two sets, the *intersection*() method returns the items common to two sets, and the *difference*() method returns a set that is the difference between two sets.

```
>>> x = {1, 2, 3, 4, 5, 6, 7}
>>> x.pop()
1
>>> x
set([2, 3, 4, 5, 6, 7])
>>> y = {'a', 'b', 'c', 2, 3}
>>> x.union(y)
set(['a', 2, 3, 4, 5, 6, 7, 'c', 'b'])
>>> x.intersection(y)
set([2, 3])
>>> x.difference(y)
set([4, 5, 6, 7])
```

Operators

Operators are reserved keywords or characters that can be used to perform operations on values and variables. The arithmetic operators supported in Python include addition (+), subtraction (−), multiplication (*), division (/), floor division (//), modulus (%), and power (**).

For logical operations (AND, OR, and NOT), the Python operators are keywords *and*, *or*, and *not* (all in lower cases). For instance, the expression "True or False" will return *True*.

```
>>> True and True
True
>>> True or False
True
```

Python also allows us to compare two values or variables to see if they are equal (==) or different (!=), or one is greater (>) or smaller (<) than the other, or one is greater or equal to (>=) or smaller or equal to (<=) the other.

Assignment operators are used in Python to assign values to variables. = is the simplest assignment operator; for instance, "x =5" assigns value "5" to the variable "x." Python also supports compound assignment operators; for instance, "x + = 1" is equivalent to "x = x + 1" (i.e., add 1 to x and assign the result to x). We will use compound operators a lot when manipulating dynamic stimuli in PsychoPy (see Chap. 2).

Other frequently used operators in Python include identity operators (*is*, *is not*) and membership operators (*in*, *not in*). These operators help to make your code resemble a natural language. In the statements below, "1 in x" returns *True* because 1 is a member of x ([1, 2, 3]).

```
>>> x = [1, 2, 3]
>>> x is False
False
>>> 1 in x
True
>>> 3 not in x
False
```

The last thing we need to discuss is operator precedence, that is, which operators are evaluated first by the Python interpreter. For instance, in "5 + 3 * 2," the multiplication operation is carried out first. There is no need to delve into the details; the official Python documentation has a section listing the operators in ascending order of precedence.[4]

Data Conversion

The built-in *type*() function can be used to reveal the type of a number or any other Python data values. The null value *None* has its own data type, i.e., *NoneType*.

```
>>> a = 3.14
>>> type(a)
<class 'float'>
>>> type(True)
<class 'bool'>
>>> type(None)
<class 'NoneType'>
```

There are built-in Python functions for data-type conversion. For instance, the *int*() function converts a floating-point number or a numeric string into an integer. The *hex*() function converts an integer to a hexadecimal string. The *str*() function converts a number or any other object into a string.

```
>>> int(10.24)
10
>>> int('110')
110
>>> hex(255)
'0xff'
>>> str(10.338)
'10.338'
```

Control Flow

In the most straightforward format, a Python script may contain multiple statements that can be evaluated and executed line by line, one after the other, like the script shown in Fig. 1.2. However, many useful scripts will involve more than the sequential execution of statements and rely on sophisticated control flow mechanisms.

[4] https://docs.python.org/3/reference/expressions.html#operator-precedence

Before we delve into the frequently used control flow features of Python, it is worth noting that Python statements are grouped by indentation instead of curly brackets or parentheses, as in some programming languages. The code snippet shown below will sum up all integers that are smaller than 100. We will discuss the *for* loop in a later section; for now, it is sufficient to note that the statement, $x + = i$, is part of the for loop, as suggested by the indentation.

```
x = 0
for i in range(100):
    x += i
print(x)
```

if *Statement*

Often a statement is executed only when certain conditions are met (i.e., conditional execution). For instance, we would like to present the target stimulus only after a fixation cross has been on the screen for 1000 milliseconds, or we will move on to the testing stage only if the accuracy of the subject is above 80% in the practice stage. In Python, a simple conditional execution usually involves *if* statements in the following format.

```
if expression_a:
    do_somehting_a
elif expression_b:
    do_somehing_b
elif expression_c:
    do_something_c
else:
    do_something_d
```

If "expression_a" is *True*, the Python interpreter will execute "do_something_a"; if not, the interpreter will evaluate the other expressions. The *elif* (if-else) and *else* statements are optional, but they help determine if one of multiple conditions are met. In the sample script below, we first show a prompt in the shell, asking users to enter a string and press ENTER to confirm. Then, we examine the length of the string.

```
s = input('Enter a string here: ')
if len(s) < 5:
    print('Length of string: s < 5')
elif len(s) < 10:
    print('Length of string: 4 < s < 10')
else:
    print('Length of string: s > 9')
```

for *Statement*

A *for* statement is used to iterate over the elements of a sequence (e.g., string, list, tuple) or an iterable object (e.g., the return value of the *range*() function).

```
for target_list in iterable_object:
    do_something_a
else:
    do_something_b
```

When evaluating a *for* statement, the items in the "iterable_object" will be assigned to "target_list" in order. The else clause is optional; it is executed (if present) when the items in the "iterable_object" are exhausted. A break statement can be used to break out a for loop if needed.

```
iterable = zip([1, 2, 3, 4, 5], [101, 102, 103, 104, 105])
print(iterable,'\n')
for i, j in iterable:
    print(i, j)
    if i > 4:
        break
print('Variables i and j will persist after the loop is finished')
print(i, j)
```

In this code snippet, we first construct a list of tuples using the *zip()* function (see the shell output below). With the help of the for loop, we then iterate over all the tuples in the list. The "target_list" in the *for* statement contains variables *i* and *j*. On each iteration, we pass the first and second value of a tuple to *i* and *j*, respectively;

then we print *i* and *j* in the shell. Note the variables *i* and *j* are not cleared when the for loop ends.

```
[(1, 101), (2, 102), (3, 103), (4, 104), (5, 105)]
1 101
2 102
3 103
4 104
5 105
Variables i and j will persist after the loop is finished
5 105
```

while *Statement*

Statements in a *while* loop are repeatedly executed as long as the conditional expression is true. A while loop can use an optional "else" clause; the interpreter will execute the else clause when the *while* loop terminates (e.g., when the conditional expression is false). A break statement can be used to exit a while loop if needed.

```
while conditional_expression:
    do_something_a
else:
    do_something_b
```

In the code snippet below, the conditional expression in the "while" statement is "i < 5." On each iteration, the variable *i* is incremented by 1 and then printed out in the shell. When the conditional expression is evaluated to *false*, e.g., when the value of *i* is 5, the loop terminates, and the *else* part prints out a notification (i.e., "while loop ended").

```
i = 0
while i < 5:
    i += 1
    print(i)
else:
    print('while loop ended')
```

The shell output from the *while* part is a number sequence from 1 to 5.

```
1
2
3
4
5
while loop ended
```

More on Looping

The *items*() method of a dictionary is iterable and can be used to loop over all key-value pairs.

```
dict = {'a':1, 'b':2, 'c':3}
for k, v in dict.items():
    print(k, v)
```

The *enumerate*() function allows us to loop over and index the elements of a sequence, such as a list or tuple. The indices start from 0 by default, but the starting index can be specified (e.g., 1 in the code snippet below).

```
s = ('a', 'b', 'c')
for i, v in enumerate(s, 1):
    print(i, v)
```

The *zip*() function is another handy tool when you need to loop over two or more sequences simultaneously.

```
for k, v in zip(('a', 'b', 'c'), (1, 2, 3)):
    print(k, v)
```

The *range*() function offers yet another way of looping.

```
for i in range(1,5):
    print('This is item: ', i)
```

List Comprehension

List comprehension is a concise way of creating one list from another list. For instance, the *range*() function can be used to create a list of integers; *list(range(5))* will return [0, 1, 2, 3, 4]. What if we would like to create a list of even numbers that are smaller than 20? One cumbersome approach is to declare an empty list and then use a loop to add in each of the even numbers.

```
evens = []
for i in range(20):
    if i%2 == 0:
        evens.append(i)
```

With list comprehension, this problem can be solved with just one line of code.

```
evens = [i for i in range(20) if i%2 == 0]
```

As shown in the code above, a list comprehension consists of brackets containing an expression (*i*) followed by a *for* statement and then zero or more *for* or *if* statements. Here is a more complicated one.

```
>>> [(x, y) for x in [1, 2] for y in [1, 2, 3, 4] if x == y]
[(1, 1), (2, 2)]
```

With list comprehension, functions can also be applied to a list to create a new list. In the statement below, we apply the *abs*() function to a number sequence (−5 to 4) generated by the *range*() function.

```
>>> [abs(x) for x in range(-5, 5)]
[5, 4, 3, 2, 1, 0, 1, 2, 3, 4]
```

Set comprehension and dictionary comprehension are also supported in Python, though the returning values are sets and dictionaries instead. The dictionary comprehension example below swaps the keys and values of dictionary *y*.

```
>>> y = {'a':1, 'b':2, 'c':3}
>>> {v:k for k, v in y.items()}
{1: 'a', 2: 'b', 3: 'c'}
```

Functions

Functions are handy when the same task needs to be performed multiple times in a script. Python comes with lots of built-in functions, such as *int*(), *list*(), *abs*(), etc. We can also define custom functions to add additional features that are useful in our experimental script. For instance, drawing a Gabor on the screen is something worth implementing in a function, so we can easily vary its position, size, spatial frequency, etc., each time we want to draw it. A function definition starts with the keyword *def*, followed by a function name and a list of arguments passed to the function. The arguments are essentially variables fed to the function for processing (e.g., Gabor size, position, and spatial frequency). The body of a function usually contains one or more statements and an optional line specifying the value that the function would return (output).

```
def function_name(arg_1, arg_2, ...):
    """Docstring"""
    Statement 1
    Statement 2
    ...
    return value
```

In Python, the return value of a function is optional; if no return value is specified, the function will return *None*. For a function that adds up two numbers, the function definition could be something like:

```
def my_add(number_1, number_2):
    """ Add up two numbers and return the sum. """
    my_sum = number_1 + number_2
    return my_sum
```

The text included in the triple quotes is a docstring, i.e., the help information for a function. For instance, *help(my_add)* will return "`Add up two numbers and return the sum.`" We call the *my_add()* function by passing two numbers to it.

```
my_add(100, 200)
```

Output

For the output of a program, we could use the *print*() function to present it on a standard output device (screen) or write it to files.

Output Formatting

It is usually helpful to format the output in a human-readable format. The "%" operator used to be the standard way of embedding values into a string. For instance, in the command below, three numbers are separated by "tabs" (\t). The "%d" are placeholders that will be replaced by the values in the tuple that follows the string, i.e., (1, 2, 3).

```
>>> print('%d\t%d\t%d' %(1, 2, 3))
1	2	3
```

With floating numbers, one can also specify how many decimal points to have in the formatted string with the % operator. For instance, to retain three decimal points, one can use "%.3f," i.e., formatting a float value ("f") with three decimal places.

```
>>> print('PI = %.3f' % 3.1415926)
PI = 3.142
```

This string formatting method still works in Python 3; however, the recommended method now is the *format*() function, which offers more flexibility. For instance, the values that we would like to include in a string do not necessarily need to appear in sequence. In the statement below, the second value "eggs" appears first, as the first {} expression is referring to the second item in the value list.

```
>>> '{1} and {0}'.format('spam', 'eggs')
'eggs and spam'
```

Keywords can also be used to format a string. In the example statement below, "Chinese food" is represented by the keyword "food" in the formatted string.

```
>>> '{food} is {adjective}.'.format(food='Chinese food', adjective='great')
'Chinese food is great.'
```

One can also use f-strings (strings prefixed by “f” or “F”) to embed the return value of Python expressions inside a string. In the example below, the Python expression is included in the { }. The value we would like to include in the string is *math.pi*. The part following the colon specifies the format of the value; the integer 10 means the value occupies 10-character spaces; “.3f” requires Python to display three decimal places for the floating-point value *math.pi*. F-string was introduced in Python from version 3.6, and it is quickly becoming popular among programmers. You will see lots of f-strings in the example scripts accompanying this book.

```
>>> f'Pi is close to {math.pi:10.3f}'
'Pi is close to      3.142'
```

Files

An experimental task usually involves manipulating files, e.g., reading task parameters from a file or writing research data collected from a testing session to a file. To open a file, the simplest solution is to call the *open*() function, which returns a file object that we can manipulate, such as adding a new data line. The *open*() function takes two arguments, a file name and a mode flag.

```
open(filename, mode)
```

The first argument is a string specifying the filename (including the path), whereas the second argument is a string specifying the operation mode. The frequently used operation modes are listed below. The mode argument is optional; *'r'* will be assumed if the mode argument is omitted:

'r'—read-only, file cannot be modified.
'w'—write, overwrite the existing file with the same filename.
'a'—appending, append to the existing file with the same filename.
'r+'—read and write.

Files usually operate in “text” mode, in which character strings are read from or written to the file. To write strings to files, use the *write()* method. To read the contents of a file, use the *read*(size) method. The *size* argument specifies the quantity of data to be read (e.g., the number of characters to read). The entire file will be read into memory if you omit this option. You can also call the *readline*() method to extract a single line from a file. Calling *readline*() repeatedly will eventually reach the end of a file. In the example script below, we first open a file called “file_op.txt” and then write three lines into it, after which the file is closed. Then, we open the file and read its contents with *readline*() and *read* () and then print the retrieved texts to the standard output (screen).

```
#! /usr/bin/env python3
#
# Filename: file_operation.py
# Author: Zhiguo Wang
# Date: 2/3/2020
#
# Description:
# Open, read, and write plain text files
file_name = 'file_op.txt'
# Open a file with 'write' permission,
# Write three lines of texts into the file, then
# close the file
print('Wring three lines of text to %s' % file_name)
file = open(file_name, 'w')
for i in range(3):
    file.write('This is line %d\n' % (i+1))
file.close()
# Open file_op.txt in read-only mode, read the first line with the
# readline() method, then close the file
print('\n\nRead the first line in %s' % file_name)
file = open('file_op.txt', 'r')
line = file.readline()
print(line)
file.close()
# Open file_op.txt in read-only mode, read four characters
print('\n\nRead the first four characters in %s' % file_name)
file = open('file_op.txt', 'r')
txt = file.read(4)
print(txt)
file.close()
# Open file_op.txt in read-only mode, then loop over all lines
file = open('file_op.txt', 'r')
for line in file:
    print(line)
file.close()
# Open file_op.txt in in a "with" statement
with open('file_op.txt', 'r') as file:
    for line in file:
        print(line.rstrip())
```

A faster and more memory-efficient way for reading lines from a file is to loop over the file object (see the code snippet below). We can also open a file in a "with" statement, ensuring the file properly closes once the file operations are done.

Note the *readline*() function will extract a line and place the "newline" character (\n) at the end. The "newline" character is necessary for the Python interpreter to tell if it has reached the end of a file, as the last line of a file does not end with the "newline" character. You can strip off the trailing "newline" character with the string method *rstrip()*.

Modules

Modules are files that you can import into a script to access the functions, classes, and constants defined within it. A module can be a script that contains custom routines you frequently call in your experiments. It usually ends with the ".py" extension, and the module name is accessible as a global variable "__name__" (two underscores).

```
>>> import pygame
>>> pygame.__name__
'pygame'
```

In Python, everything (constants, functions, etc.) is an object. The objects in a module are accessible with the module name, followed by a period (".") and the name of the object. To access the objects in a module, you need to import the module into your script or the shell with the *import* command. You can import a module and rename it.

```
>>> import numpy as np
```

You can import all objects from a module (generally not recommended).

```
>>> from pygame import *
```

You can also selectively import a few objects from a module.

```
>>> from math import sin, cos, hypot
```

Choose a Python Module to Build Your Experiments

A computer-based experimental task can be thought of in some respects as a video game. Visual and auditory stimuli are presented, and the participants respond to various task manipulations with a keyboard, a mouse, or a gamepad. Several Python modules have been developed for vision scientists, for example, PsychoPy, pyexperiment, and visionEGG. Those Python modules feature carefully implemented functions for computer-based experiments. This book will focus on PsychoPy, as it is widely used and has many attractive features (Chap. 2). We will also briefly introduce Pygame, which works well as a lightweight solution for some studies (Chap. 3).

This book will cover tools and functions that are essential for building experimental tasks. For a comprehensive introduction to Pygame, I would recommend *Beginning Game Development with Python and Pygame: From Novice to Professional* by Will McGugan. For PsychoPy, consider *Building Experiments in PsychoPy* by Jon Pierce. It is a good reference for the graphical programming interface of PsychoPy, but it also covers coding in chapters targeting more advanced users.

Chapter 2
Building Experiments with PsychoPy

Contents

There are a few Python modules or Python-based tools that can be used to build psychological experiments. PsychoPy, which also works as a standalone application, is one of the most popular. The low-level libraries that PsychoPy heavily depends on include Pyglet, Pygame, and a few other libraries for image manipulation, movie and audio playback, etc. Building upon those libraries is a set of convenient functions for generating complex visual stimuli (e.g., checkerboard and Gabor patches), registering keyboard and mouse events, and interfacing with eye trackers, EEG, and other frequently used research equipment. PsychoPy also includes handy tools for monitor calibration, color conversion, etc.

As an actively maintained and rapidly evolving research tool, it is inevitable that some of PsychoPy's features will become deprecated and other new features will become available. This chapter will cover the essential functions that were active

Z. Wang, *Eye-Tracking with Python and Pylink*,
https://doi.org/10.1007/978-3-030-82635-2_2

when this book was written and are likely to remain active in the immediate future. I will first introduce a few core modules of PsychoPy (e.g., visual, monitors, events), with examples, and then present a simple method for handling experimental trials.

Installing PsychoPy

PsychoPy can be installed as a Python module with "pip." For instance, executing "py -3.6 -m pip install psychopy" from the command prompt will install PsychoPy and all its dependencies for Python 3.6 on Windows. If a graphical user interface is preferred, you can install the standalone version of PsychoPy on Windows or macOS, freely available from https://www.psychopy.org. The standalone version comes with Python 3.6. It is a self-contained Python environment that does not interfere with any other versions of Python installed on your computer.

The standalone version of PsychoPy has a graphical programming interface—known as "Builder"—for building experimental tasks through dragging and dropping functional components onto a timeline. This book will not cover the Builder interface. Instead, we will focus on the "Coder" interface of PsychoPy, which allows users to write and debug scripts as if PsychoPy is just another Python IDE. The PsychoPy Coder interface consists of an editor, a Python shell, and an Output window for presenting debugging information, e.g., warnings, errors, and the output of a script (see Fig. 2.1)

At the time this book was being written, the PsychoPy version was 3.2 (https://www.psychopy.org/). Nevertheless, the example scripts included in this book should also work in newer versions.

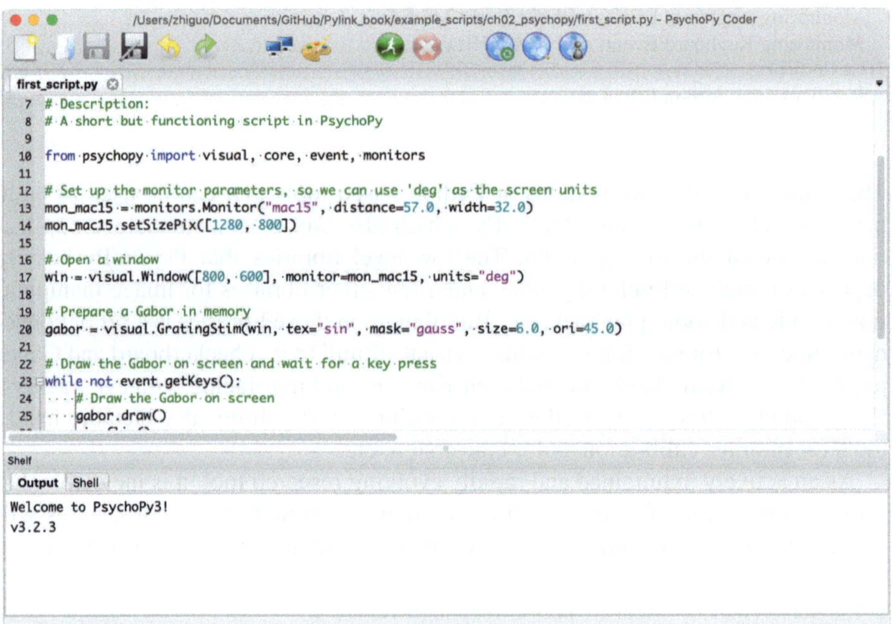

Fig. 2.1 The Coder interface of PsychoPy.

A Basic PsychoPy Script

Before we discuss the various PsychoPy modules in detail, we will first go through a short script to get familiar with the essential elements of PsychoPy. This short script will show a drifting Gabor on the screen until a key is pressed. To run this script, open it in PsychoPy Coder, and then click the Run button on the menu bar or press the hotkey F5.

```
#!/usr/bin/env python3
#
# Filename: first_script.py
# Author: Zhiguo Wang
# Date: 2/7/2021
#
# Description:
# A short but functioning script in PsychoPy

from psychopy import visual, core, event, monitors


# Set up the monitor parameters, so we can use 'deg' as the screen units
mon_mac15 = monitors.Monitor("mac15", distance=57.0, width=32.0)
mon_mac15.setSizePix([1280, 800])

# Open a window
win = visual.Window([800, 600], monitor=mon_mac15, units="deg")

# Prepare a Gabor in memory
gabor = visual.GratingStim(win,   tex="sin",   mask="gauss",   size=6.0,
                           ori=45.0)

# Draw the Gabor on screen and wait for a key press
while not event.getKeys():
    # Draw the Gabor on screen
    gabor.draw()
    win.flip()

    # Update the phase following each screen refresh
    gabor.phase += 0.05

# close the window and quit PsychoPy
win.close()
core.quit()
```

The structure of this script is pretty straightforward. We first import the PsychoPy modules needed for this script, notably the *visual, core, event*, and *monitor* modules. Then, we create a Monitor object with *monitor.Monitor*() to store the various monitor parameters, such as eye-to-screen distance (in cm), screen width (in cm), and the native screen resolution (in pixels).

```
mon_mac15 = monitors.Monitor("mac15", distance=57.0, width=32.0)
mon_mac15.setSizePix([1280, 800])
```

With the correct monitor parameters, we can then open a Window with *visual.Window*(). We will discuss the frequently used parameters of *visual.Window()* in detail in a later section. Here we specify the dimension of the screen (in pixels), the configuration for the monitor, and the screen units we will use to control stimulus size.

```
win = visual.Window([800, 600], monitor=mon_mac15, units="deg")
```

One of the advantages of PsychoPy is that it allows you to manipulate the properties of the stimuli (e.g., position, color, spatial frequency, phase, etc.) on the fly during testing. In this script, we first create a Gabor patch with *visual.GratingStim()*. We set its size to 6.0° of visual angle (we will discuss screen units later) and orientation to 45.0°.

```
gabor = visual.GratingStim(win, tex="sin", mask="gauss", size=6.0,
                           ori=45.0)
```

We call the *draw*() method to prepare the Gabor in the video memory; then, we call the *flip*() function to show it on the screen.[1] We repeat these operations in a *while* loop and update the phase of the Gabor (by adding 0.05) following each screen refresh. So, the Gabor appears drifting on the screen.

```
while not event.getKeys():
    # Draw the Gabor on screen
    gabor.draw()
    win.flip()

    # Update the phase of the Gabor following each screen refresh
    gabor.phase += 0.05
```

[1] Modern video graphics cards all support hardware double buffering. At any time, the contents in the front buffer are actively being displayed on the screen, while the back buffer is being drawn. When the drawing is complete in the back buffer, the two buffers can be switched (flipped) so the back buffer becomes the front buffer and its content is shown on screen, usually in sync with the monitor vertical blanking signal.

The *event.getKeys*() function will return a list of the keys you have pressed since the last call to this function. We use "*not event.getKeys*()" as the conditional expression in the while loop. The while loop will continue as long as the list (of keys) returned by event.*getKeys*() is empty. If the list is not empty (e.g., if a key has been pressed), we break the loop and close the PsychoPy window.

This short script requires the *visual*, *core*, *event*, and *monitors* modules of PsychoPy. The *visual* module includes routines for window management and visual stimulus generation; the *core* module contains basic functions related to timing (e.g., clock). The *event* module handles keyboard and mouse events, whereas the *monitors* module provides tools for monitor configuration. Depending on the nature and complexity of your experimental script, you may need to use other PsychoPy modules, such as *hardware*, *data*, *sound*, and *gui*. This chapter will focus on modules critical to eye-tracking studies, notably those related to visual stimulus generation and keyboard/mouse event registration. For an overview of PsychoPy and its various features, please refer to Pierce et al. (2019) and Pierce and Macaskill (2018).

Opening a Window

A PsychoPy window is essentially a container for your visual stimuli, and opening a window is usually the first step in an experimental script. Lots of parameters can be configured when you call *visual.Window*() to open a PsychoPy window. It is usually safe to take the default parameter values, with a few exceptions, such as *size*, *color*, *fullscr*, and *units*. Detailed information about all the parameters is listed in the PsychoPy documentation. Here we will highlight a few.

```
visual.Window(size=(800, 600), pos=None, color=(0, 0, 0), colorSpace='rgb',
rgb=None, dkl=None, lms=None, fullscr=None, allowGUI=None, monitor=None,
bitsMode=None, winType=None, units=None, gamma=None, blendMode='avg',
screen=0, viewScale=None, viewPos=None, viewOri=0.0, waitBlanking=True,
allowStencil=False, multiSample=False, numSamples=2, stereo=False,
name='window1', checkTiming=True, useFBO=False, useRetina=True,
autoLog=True, gammaErrorPolicy='raise', bpc=(8, 8, 8), depthBits=8, sten-
cilBits=8, backendConf=None)
```

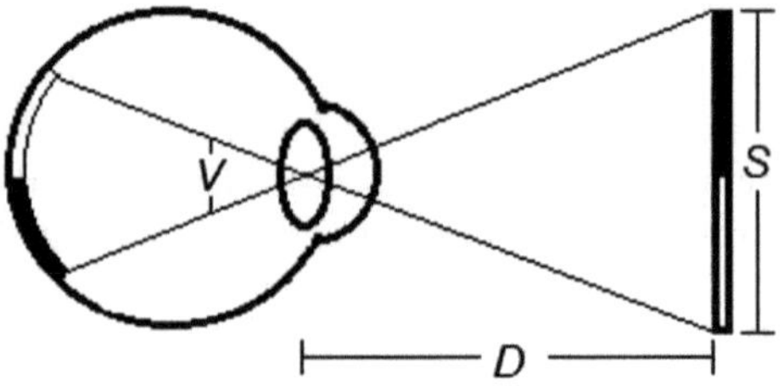

Fig. 2.2 Visual angle calculation. In this illustration, the visual angle (V) for object S is 2 * arctan(S/2D)

Screen Units

The screen unit in computer graphics is a "pixel." The pixel coordinate of the top-left corner of the screen is usually (0, 0); the bottom-right corner of the screen corresponds to the maximum number of horizontal and vertical pixels minus 1. One departure of PsychoPy from this convention is that the origin of the screen coordinates is the center of the screen, with negative values running leftward and downward and positive values running rightward and upward, as in the Cartesian coordinates.

In addition to screen pixels, PsychoPy also supports several other screen units. The most useful screen unit in vision science is "visual angle" (in degrees), which reflects the size of an object on the retina (see Fig. 2.2 for an illustration).

Pixels are the native units in computer graphics. To draw a line that is 1° (visual angle) long on the monitor, we need to know how many pixels correspond to 1° of visual angle. To get this info, we need the resolution of the screen (in pixels), the physical size of the visible screen area (in centimeters), and the eye-to-screen distance (in centimeters). Things can be a bit more complicated if we consider that most computer monitors have flat screens, so the eye-to-screen distance varies with gaze position. In PsychoPy, you can use one of the following visual angle units:

- *deg*—assumes 1 degree of visual angle spans the same number of pixels at all parts of the screen
- *degFlatPos*—corrects screen flatness only for stimuli position, no change in stimulus size and shape
- *degFlat*—corrects screen flatness for each vertex of the stimuli so that a square will get larger and rhomboid in the peripheral

In addition to degrees of visual angle, PsychoPy also supports the following screen units:

- *norm*—normalized, visible screen area ranges from −1 to 1 in both x and y.
- *cm*—centimeters on screen.
- *height*—relative to the height of the screen. The vertical screen range is −0.5 to 0.5; for a 4:3 ratio screen, the horizontal screen range is −0.67 to 0.67.
- *pix*—screen pixel.

Screen pixel ('pix') is what I would recommend for eye-tracking studies, as the gaze position returned by the tracker is in screen pixel coordinates.

Monitor

To correctly present stimuli in the required screen unit (e.g., degree of visual angle), PsychoPy requires some information about the monitor, for instance, the viewing distance and the physical size and resolution of the screen. These parameters can be configured in the PsychoPy Coder interface, but I strongly recommend defining a *Monitor* object to specify these parameters explicitly. This practice improves code transparency.

In the code snippet below, we create a *Monitor* object, name it as "*mac15,*" set the screen resolution to 1280 × 800 pixels, and set the screen width and viewing distance to 32 and 57 cm, respectively. We then pass this *Monitor* object to the "monitor" parameter of *visual.Window*() when opening a new PsychoPy window.

```
mon_mac15 = monitors.Monitor("mac15", distance=57.0, width=32.0)
mon_mac15.setSizePix([1280, 800])
win = visual.Window([800, 600], monitor=mon_mac15, units="deg")
```

Gamma

In computer graphics, the output device (e.g., a monitor or printer) does not always give the requested output. For instance, the RGB values for black and white are (0, 0, 0) and (255, 255, 255). You may imagine that increasing the values in the RGB triplets from 0 to 255 will give you a grayscale of linearly increasing brightness. As is clear from Fig. 2.3, this is not the case; the resulting grayscale on the screen is not linear but follows a gamma function (in the simplest format, $y = x^{\gamma}$).

The nonlinearity in video output is usually not a big issue in many experimental tasks, but it can easily mess up a psychophysics task that measures perceptual thresholds. PsychoPy has provided functions for deriving gamma value from measured screen luminance; please see the *monitors* section of the PsychoPy documentation for details.

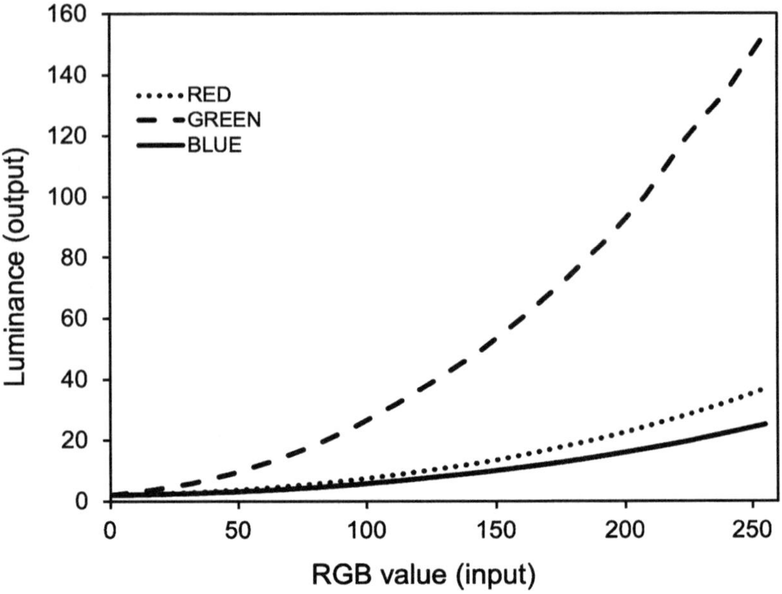

Fig. 2.3 The nonlinear output of computer monitors generally follows the gamma function. This illustration shows how the screen luminance, measured with a photometer, varies with the requested RGB values

Vertical Blanking

Old-fashioned cathode ray tube (CRT) monitors displayed an image by scanning a beam of electrons across the screen in a pattern of horizontal lines. At the end of each line, the beam returns to the start of the next line (horizontal retrace); at the end of the last line, the beam returns to the top of the screen (vertical retrace). The monitor blanks the beam to avoid displaying a retrace line during horizontal and vertical retraces (see Fig. 2.4 for an illustration). The refresh rate of a monitor describes precisely how fast it draws a full screen. A 60 Hz monitor will redraw the screen 60 times a second; so, to draw a full screen, the monitor takes 1000/60 = ~16.67 ms.

The implication is that the pixels of an image will not show up altogether, but instead, they appear on screen one by one. We can request the monitor to draw an image any time we want, but it is best to wait until the next vertical retrace. Otherwise, you may see screen tearing—where the "lower" portion of an object is drawn before its "upper" portion. Modern liquid crystal displays (LCDs) no longer rely on a beam of electrons, but many LCD monitors still update the screen in a similar fashion.

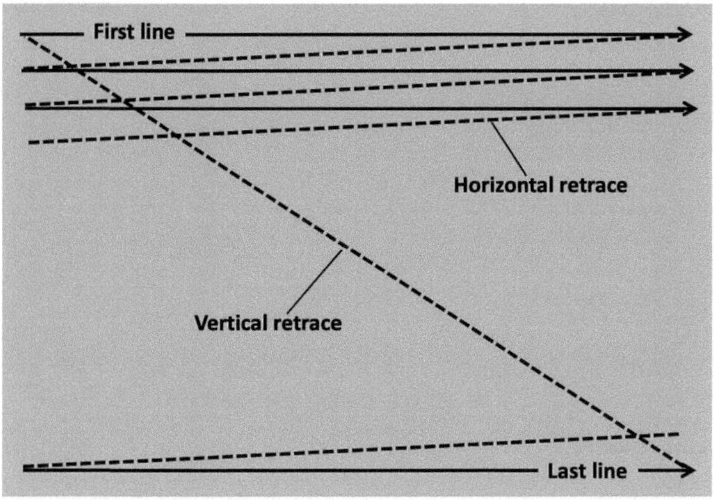

Fig. 2.4 Horizontal and vertical retraces

Visual stimuli are first drawn in a memory buffer, and we need to call the *flip*() function to show the contents in the buffer on the screen. Setting the "waitBlanking" parameter to *True* ensures that the *flip*() function does not execute until a vertical blanking signal is received. So, the time returned by *flip*() is the actual start time of a new screen.

Call a Function on Window Flip

One handy method of a window object in PsychoPy is *callOnFlip*(), i.e., execute something at the same time we *flip*() the window. For instance, we can use this method to send out a TTL signal (see Chap. 7) to mark the onset time of a visual stimulus in EEG data.

The first parameter for *callOnFlip*() is a function to be executed on the upcoming window flip. The other parameters are the ones you would normally pass directly to this function. The example script below first defines a function that prints out the current time, i.e., *print_time()*. Then, we request PsychoPy to execute this function on the upcoming window flip. After the window flip, we print out the current time to check if the execution time of the *print_time()* function is the same as the window flipping time. The two timestamps should match as the *print_time()* function takes little time to return.

```
#!/usr/bin/env python3
#
# Filename: demo_callOnFlip.py
# Author: Zhiguo Wang
# Date: 2/7/2021
#
# Description:
# Check the timing accuracy of the .callonFlip() function

from psychopy import visual, core

win = visual.Window(size=[1200, 800], units="pix", fullscr=True)

# A function to print out the current time
def print_time():
    current_t = core.getTime()
    print(f'print_time() executed at time: {current_t:.3f}')

# In a for-loop, check if print_time() is executed at the same time as
# the window flip
for i in range(10):
    win.callOnFlip(print_time)
    flip_t = win.flip()
    print(f'Actual window flipping time was: {flip_t:.3f}')
    core.wait(0.5)

# Close the window and quit PsychoPy
win.close()
core.quit()
```

Taking Screenshots

A PsychoPy visual.*Window()* object also has a few other useful methods. For instance, the *fps*() method returns an estimate of the frame rate since the last call to this function or since the window was initially opened. The *getActualFrameRate*() method will run a test and report the frame rate of the screen. You can also capture the screen contents as images or videos with the *getMovieFrame*() and *saveMovieFrames*() methods. The short script below illustrates these features. Taking a screenshot can be time-consuming, and it is not recommended for timing critical tasks.

```
# Filename: demo_screenshot.py
# Author: Zhiguo Wang
# Date: 11/7/2020
#
# Description:
# Get the actual frame rate, then take a screenshot

from psychopy import visual, core


# Open a window
win = visual.Window(size=[800, 600], units="pix")

# Get frame rate (frame per second)
fps = win.getActualFrameRate()
print(f'Frame rate is: {fps} FPS')

# Capture the screen
win.color = (0, 0, 0)
win.getMovieFrame()

# Show the screen for 1.0 second
win.flip()
core.wait(1.0)

# Save captured screen to a JPEG
win.saveMovieFrames("gray_window.jpg")

# Quit PsychoPy
win.close()
core.quit()
```

The output of the above script revealed that the actual frame rate of my MacBook is 60 FPS (frame per second).

```
Frame rate is: 60.006520908485754 FPS
```

Visual Stimuli

One major strength of PsychoPy is the variety of visual stimuli it supports, which include shapes, images, texts, and patterns (e.g., checkerboard). This section will briefly introduce some of the more frequently used stimuli types.

Shapes

A variety of shapes can be drawn in PsychoPy, for example, rectangles, circles, polygons, and, of course, lines. In addition to dedicated drawing functions such as *visual.Rect*(), *visual.Circle*(), *visual.Polygon*(), and *visual.Line*(), PsychoPy also has an abstract class called *visual.ShapeStim*(), which is capable of drawing shapes with arbitrary numbers of vertices.

```
visual.ShapeStim(win, units='', colorSpace='rgb', fillColor=False,
lineColor=False, lineWidth=1.5, vertices=((- 0.5, 0), (0, 0.5), (0.5, 0)),
windingRule=None, closeShape=True, pos=(0, 0), size=1, ori=0.0, opac-
ity=1.0, contrast=1.0, depth=0, interpolate=True, name=None, autoLog=None,
autoDraw=False, color=False, lineRGB=False, fillRGB=False,
fillColorSpace=None, lineColorSpace=None)
```

The code below shows how to draw both lines and rectangles with *visual. ShapeStim*(); for lines, the "*closeShape*" parameter needs to be *False*. The other shape functions are all special cases of *visual.ShapeStim*().

```
#!/usr/bin/env python3
#
# Filename: shapes.py
# Author: Zhiguo Wang
# Date: 2/7/2021
#
# Description:
# Drawing in PsychoPy

from psychopy import visual, event, core
```

```
# Open a Window
win = visual.Window(size=[800, 600], units='pix')

# Line
line_vertices = [(-400, -300), (400, 300)]
line = visual.ShapeStim(win, vertices=line_vertices,
lineColor='white', closeShape=False)

# Rectangle
rect_vertices = [(-400, -300), (-320, -300), (-320, -240), (-400, -240)]
rect = visual.ShapeStim(win, vertices=rect_vertices,
fillColor='blue', lineWidth=0)

# Polygon
poly = visual.Polygon(win, edges=6, radius=100, fillColor='green')

# Move the rectangle along the line and change the filling color of
# the polygon when it overlaps with the rectangle
while True:
    if rect.overlaps(poly):
        poly.fillColor = 'red'
    else:
        poly.fillColor = 'green'
    line.draw()
    poly.draw()
    rect.draw()
    win.flip()
    # Update the position of the rectangle following each flip
    rect.pos += (4, 3)
    # Break out when the rectangle reaches the end of the line
    if rect.contains((400, 300)):
        break

# Close the window and quit PsychoPy
win.close()
core.quit()
```

The short script above will continuously update the position of the rectangle so it moves along the line we draw on the screen. You can also, of course, dynamically change its color, size, etc., if needed.

```
rect.pos += (4, 3)
```

One useful feature of *visual.ShapeStim*() is the *contains()* method, which checks if a screen point (x, y) is inside the border of a shape. In the example script above, we use *contains()* to check if the top-right corner of the screen is inside the rectangle; if so, we terminate the task.

As noted above, there are dedicated methods such as *visual.Rect*(), which can be used to draw various shapes in PsychoPy. In the example script, we draw a hexagon at the center of the screen with *visual.Polygon*(), in which we specify the number of edges in a polygon (6) and the radius.

```
poly = visual.Polygon(win, edges=6, radius=100, fillColor='green')
```

The *overlaps*() method is another useful feature of the *visual.Shape*() class. It allows you to check if one shape overlaps with another. In the example script, we check if the rectangle overlaps with the hexagon. If so, we change the color of the hexagon to 'red'.

```
if rect.overlaps(poly):
    poly.fillColor = 'red'
else:
    poly.fillColor = 'green'
```

GratingStim

visual.GratingStim() is one of the most versatile visual stimuli in PsychoPy. It is a texture behind an optional transparency mask (also known as a filter). Both the texture and mask can be arbitrary bitmaps that repeat (cycle) either horizontally or vertically, for instance, gratings, Gabor patches (gratings with a Gaussian mask), and checkerboards.

```
visual.GratingStim(win, tex='sin', mask='none', units='', pos=(0.0, 0.0),
size=None, sf=None, ori=0.0, phase=(0.0, 0.0), texRes=128, rgb=None,
dkl=None, lms=None, color=(1.0, 1.0, 1.0), colorSpace='rgb', contrast=1.0,
opacity=None, depth=0, rgbPedestal=(0.0, 0.0, 0.0), interpolate=False,
blendmode='avg', name=None, autoLog=None, autoDraw=False, maskParams=None)
```

Here we will briefly introduce some of the frequently used parameters and methods, of course, with examples. The texture ("tex") parameter could be *'sin'* (sine wave), *'sqr'* (square wave), *'saw'* (sawtooth wave), *'triangle wave'*, or *None* (default). The texture can also be an image with power-of-two dimensions (e.g., 128 × 128) or a NumPy array (https://numpy.org) with values ranging between −1 and 1. The "mask" parameter could be set to *'circle'*, *'gauss'* (2D Gaussian filter), *'raisedCos'* (raised-cosine filter), *'cross'*, or *None*. For a detailed discussion on texture and mask in PsychoPy, I would recommend an excellent tutorial by Sebastiaan Mathot.[2]

Another critical parameter for visual.*GratimgStim()* is spatial frequency ("sf"). We should consider the screen units when setting the "sf" parameter. If the screen unit is pixel ('pix'), the spatial frequency has to be a fraction of 1, e.g., *sf* = 1/20.0, as there is no way to repeat a bitmap within a single pixel.

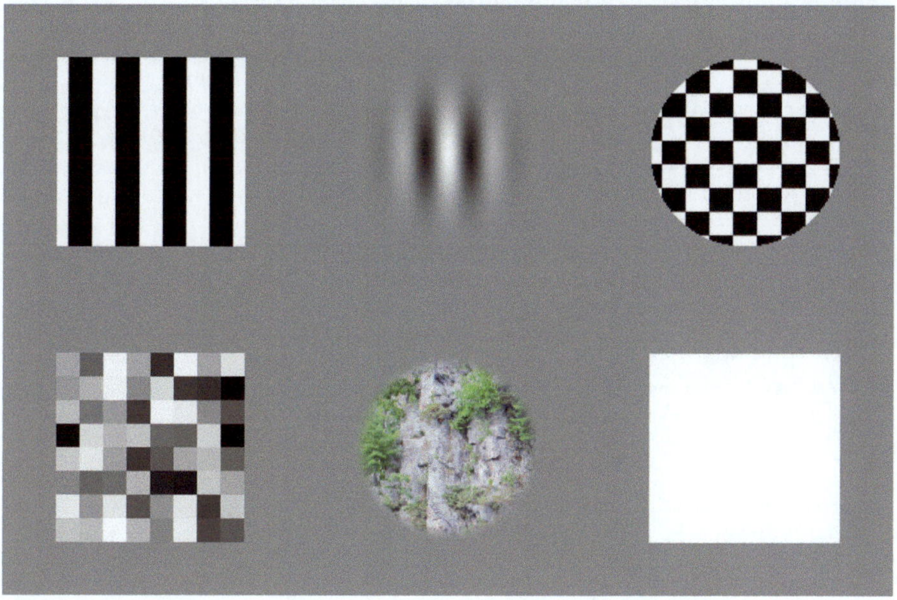

Fig. 2.5 visual.GratingStim() can be used to generate various visual patterns

[2] http://www.cogsci.nl/blog/tutorials/211-a-bit-about-patches-textures-and-masks-in-psychopy.

```
#!/usr/bin/env python3
#
# Filename: demo_GratingStim.py
# Author: Zhiguo Wang
# Date: 2/7/2021
#
# Description:
# The GratingStim() function in PsychoPy

from psychopy import visual, core
import numpy as np

# Open a window
win = visual.Window(size=(600, 400), units="pix", color=[0, 0, 0])

# Prepare the stimuli in memory
grating = visual.GratingStim(win, tex='sqr', mask=None,
                             size=128, sf=1/32.0, pos=(-200, 100))
gabor = visual.GratingStim(win, tex='sin', mask='gauss',
                           size=128, sf=1/32.0, pos=(0, 100))
checker = visual.GratingStim(win, tex='sqrXsqr', mask='circle',
                             size=128, sf=1/32.0, pos=(200, 100))
# Customize texture
# a 8 x 8 grid of random values between -1 and 1
custom_texture = np.random.random((8, 8))*2 - 1
numpy_texture = visual.GratingStim(win, tex=custom_texture, mask=None,
                                   size=128, pos=(-200, -100))
# Use an image as the texture
image_texture=visual.GratingStim(win,tex='texture.png',mask='raisedCos',
                                   size=128, pos=(0, -100))
# You get a rectangle with no texture or mask
no_texture = visual.GratingStim(win, tex=None, mask=None,
                                size=128, pos=(200, -100))

# Show the stimuli
grating.draw()
gabor.draw()
checker.draw()
numpy_texture.draw()
image_texture.draw()
no_texture.draw()
win.flip()

# Take a screenshot and save it to a PNG
win.getMovieFrame()
```

```
win.saveMovieFrames('gratings.png')

# Show the stimuli for 5 seconds, then close the window and quit PsychoPy
core.wait(5.0)
win.close()
core.quit()
```

The example script sets the texture to a square wave, a sine wave, a checkerboard, an image, and an 8 × 8 NumPy array with random values ranging between −1 and 1. The mask is set to None, "gauss," "circle," or "raisedCos." If both the texture and mask parameters were both *None*, the result is a bright square, as the default contrast is 1 (maximum). You will need a 128 × 128 (pixel) PNG image file named "texture.png" to run this script. Please see Fig. 2.5 for the resulting screen.

TextStim

Presenting text on the screen is not as trivial as it sounds. While it is relatively simple to show text messages to the subjects, correctly formatting text for a reading task in PsychoPy can be challenging. For instance, in an eye-tracking task, it is preferable to segment the text into interest areas to facilitate later data analysis. *visual.TextStim*() or *visual.TextBox*() can be used to present text stimuli, but text rendering in PsychoPy is generally slow. Dynamically updating the text during testing may not be a good idea in certain scenarios. Here we briefly discuss a few parameters of *visual.TextStim*(). For an example script, please see the *Aperture* section.

```
visual.TextStim(win, text='Hello World', font='', pos=(0.0, 0.0),
depth=0, rgb=None, color=(1.0, 1.0, 1.0), colorSpace='rgb', opacity=1.0,
contrast=1.0, units='', ori=0.0, height=None, antialias=True, bold=False,
italic=False, alignHoriz=None, alignVert=None, alignText='center',
anchorHoriz='center', anchorVert='center', fontFiles=(), wrapWidth=None,
flipHoriz=False, flipVert=False, languageStyle='LTR', name=None,
autoLog=None)
```

Font and Font Files

The "font" parameter requires a system font that will be used to display the text. Exactly which font names can be passed to *visual.TextStim*() depends on the platform. On macOS, the names of the fonts can be found with the Font Book application. On Windows, the font names can be listed by going to the Fonts section of the Control panel. There is no easy way to retrieve the names of all available system fonts programmatically. However, because the Matplotlib module is a dependency for PsychoPy, one can use *matplotlib.font_manager* to retrieve names of the system fonts. In the code snippet below, *get_fontconfig_fonts*() returns all the font files; we then use *get_name*() to retrieve the name of the fonts. Running this script from the command line should return all the font names you can use in PsychoPy.

```
#!/usr/bin/env python3
#
# Filename: get_fontNames.py
# Author: Zhiguo Wang
# Date: 2/7/2020
#
# Description:
# Retrieve the names of all available system fonts
# Run this script from the command line

from matplotlib import font_manager


f_list = font_manager.get_fontconfig_fonts()
f_names = []
for font in f_list:
    try:
        f = font_manager.FontProperties(fname=font).get_name()
        f_names.append(f)
    except:
        pass

print(f_names)
```

Right-to-Left Text

You can only set this parameter when initializing the *TextStim()* object. This parameter helps to correctly display texts from some languages that are written right to left (e.g., Hebrew). The default value for this parameter is '*LTR*' (left to right, case sensitive); setting it to '*RTL*' will correctly display text in right-to-left languages.

Wrap Width

This parameter specifies the width the text should run before wrapping. If we do not set the *TextStim*() unit explicitly, the unit of the *wrapWidth* parameter will be the same as the screen.

```
text = visual.TextStim(win, text="Moving window example by Zhiguo "*24,
                        height=30, color='black', wrapWidth=760)
```

Aperture

The *visual* module of PsychoPy allows users to set up an "aperture" to hide or reveal part of a window.

```
visual.Aperture(win, size=1, pos=(0, 0), ori=0, nVert=120,
shape='circle', inverted=False, units=None, name=None, autoLog=None)
```

To use the aperture function, we need to open a window with the *allowStencil* parameter enabled.

```
win = visual.Window(size=(800, 600), units="pix", allowStencil=True)
```

Setting an aperture requires just a single line of code. The "shape" parameter is arguably the most parameter; it can be "circle," "triangle," or, more usefully, a list containing the vertices of a freehand shape. If needed, the "inverted" parameter can be set to *True* to show the contents outside the aperture rather than that inside it. In combination with a high-speed eye tracker, capable of passing gaze data back to the computer running PsychoPy, this function can be used to create a gaze-contingent mask, simulating a macular degeneration condition (see Fig. 2.6).

I used the following script to generate Fig. 2.6. Instead of an eye tracker passing gaze data, the script uses the location of the mouse to simulate gaze, and the blind spot moves with the mouse.

```
#!/usr/bin/env python3
#
# Filename: demo_aperture.py
# Author: Zhiguo Wang
# Date: 2/7/2020
#
# Description:
# Simulating a macular degeneration condition in PsychoPy

from psychopy import visual, core, event



# Open a window
win = visual.Window(size=(800, 600), units="pix", allowStencil=True)

# Create an aperture of arbitrary shape
vert = [(0.1, .50), (.45, .20), (.10, -.5), (-.60, -.5), (-.5, .20)]
apt = visual.Aperture(win, size=200, shape=vert, inverted=True)
apt.enabled = True

# Initialize a mouse object
mouse = event.Mouse(visible=False)

# Prepare the stimuli
text = visual.TextStim(win, text="Moving window example by Zhiguo "*24,
                       height=30, color='black', wrapWidth=760)

# Mouse-contingent moving window
while not event.getKeys():
    apt.pos = mouse.getPos()
    text.draw()
    win.flip()

# Close the window and quit PsychoPy
win.close()
core.quit()
```

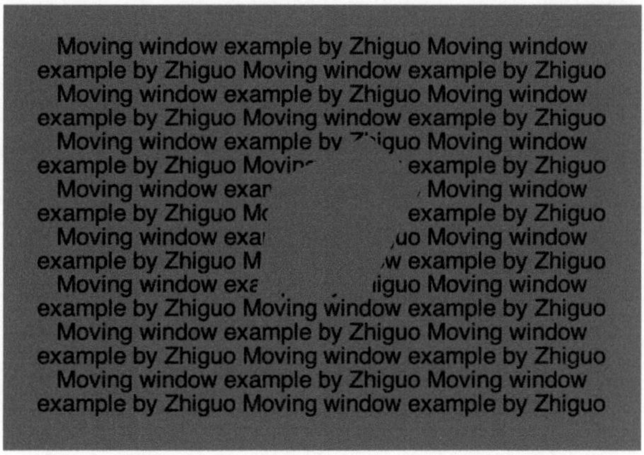

Fig. 2.6 Using aperture to simulate a macular degeneration condition

PsychoPy also supports other types of visual stimuli, such as random dots, radial stimuli (e.g., rotating wedges), rating scales, video, and many others. To learn more about these stimuli, it is worth looking over the various example scripts accessible from the PsychoPy Coder menu.

Mouse

The mouse is an indispensable interaction channel in many experimental tasks. Some researchers have even used the mouse movement trajectories to tap into the temporal dynamics of decision-making, though the timing precision of the mouse position samples can be questionable. To use a mouse in a project, first, create a mouse object with *event.Mouse*(). Then, we can access the various properties and methods of the mouse object. For instance, *getPos*() will return the current mouse position, and *setPos*() will put the mouse cursor at a given screen position.

The short script below demonstrates the most frequently used mouse functions. The task is simple: we present two options on the screen and instruct the participant to move the mouse to click one of the two options. We plot the movement trajectory while the mouse is moving.

```python
#!/usr/bin/env python3
#
# Filename: demo_mouse.py
# Author: Zhiguo Wang
# Date: 2/6/2021
#
# Description:
# A short demo illustrating the various mouse functions in PsychoPy

from psychopy import visual, event, core


# Open a window
win = visual.Window(size=(800, 600), winType='pyglet', units='pix')

# Create a Mouse object
mouse = event.Mouse(visible=True, win=win)

# Prepare the stimuli in memory
text_prompt = visual.TextStim(win=win, text='Do you like PsychoPy?',
                              height=30, pos=(0, 250))
text_yes = visual.TextStim(win=win, text='YES', height=30,
                           pos=(-200, 150), color='red')
text_no = visual.TextStim(win=win, text='NO',  height=30,
                          pos=(200, 150), color='green')
circle_yes = visual.Polygon(win=win, edges=32, radius=60,
                            pos=(-200, 150), fillColor='white')
circle_no = visual.Polygon(win=win, edges=32, radius=60,
                           pos=(200, 150), fillColor='white')
fix_cross = visual.TextStim(win=win, text='+', height=30, pos=(0, -150))
mouse_traj = visual.ShapeStim(win=win, lineColor='black',
                              closeShape=False, lineWidth=5)

# Clear cached events
event.clearEvents()

# Set the mouse position, so the movement starts from the fixation cross
mouse.setPos((0, -150))

# Use a list to store the mouse position
traj = [mouse.getPos()]

# In a while loop, check if the "yes" or "no" circle has been clicked
while not (mouse.isPressedIn(circle_no) or mouse.
isPressedIn(circle_yes)):
```

```
        # Following a position change, add the new mouse position to 'traj'
        if mouse.mouseMoved():
            traj.append(mouse.getPos())

        # Put stimuli on display and draw the mouse trajectory
        text_prompt.draw()
        circle_no.draw()
        circle_yes.draw()
        text_no.draw()
        text_yes.draw()
        fix_cross.draw()
        mouse_traj.vertices = traj  # this can be slow
        mouse_traj.draw()
        win.flip()

# Close the window and quit PsychoPy
win.close()
core.quit()
```

The script itself is relatively straightforward; the only thing worth mentioning here is the *isPressedIn*() method, which we use to check if a visual stimulus has been clicked or not. It is recommended to call *event.clearEvents*() at the beginning of the script, so cached mouse and keyboard events will not interfere with response registration. This script will draw the trajectory as you move the mouse, until you click the "YES" or "NO" response options (see Fig. 2.7).

Keyboard

PsychoPy offers multiple ways to handle keyboard events. The simplest solution is to use *event.getKeys()* and *event.waitKeys*(). The timing precision of both methods is not great. When timing is critical in an experimental task, the recommended solution is the *keyboard* module, which wraps up the PsychHID library from Psychtoolbox.

Monitoring Keyboard Events with the Event Module

The *event.waitKeys*() function halts everything while waiting for input from the keyboard. The *event.getKeys*() function, on the other hand, is usually used to check if any key is pressed in a *while* loop. Both functions require users to specify a "keyList" parameter, i.e., *keyList = ['z', 'slash']* will filter out all keys except for z and slash. The tricky thing is the names of the keys (e.g., "enter," "backslash") are not

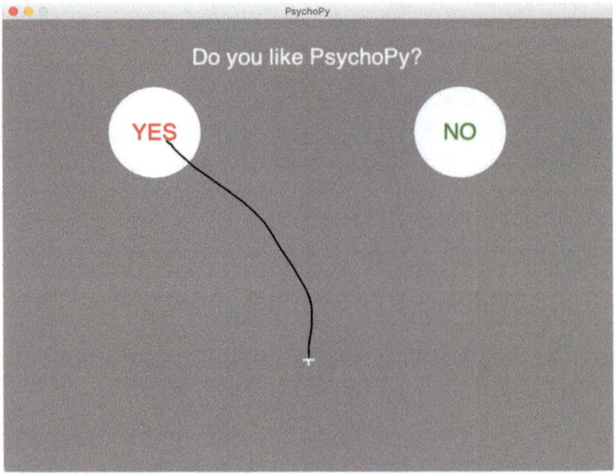

Fig. 2.7 Mouse trajectory in a decision task

documented. Nevertheless, it is trivial to figure out the key names with a few lines of code.

```
from psychopy import visual, event, core
my = visual.Window(size=(100, 100))
while True:
    key = event.waitKeys(timeStamped=True)
    print(key)
```

The shell output from this code snippet is shown in the box below. The *event.waitKeys*() function returns a list of pressed keys. Because we set the *timeStamped* parameter to *True*, all keypresses have timestamps.

```
[['a', 5.818389608990401]]
[['b', 6.602048815984745]]
[['apostrophe', 13.980140805011615]]
[['slash', 15.94004777900409]]
[['backslash', 18.18008749600267]]
[['return', 19.80443760601338]]
```

Monitoring Keyboard Events with PsychHID

Registering keyboard events with the Psychtoolbox PsychHID library is a new feature in PsychoPy. This function is only available for 64-bit Python; for 32-bit Python, it reverts to *event.getKeys*().

```
#!/usr/bin/env python3
#
# Filename: demo_PsychHID.py
# Author: Zhiguo Wang
# Date: 2/6/2021
#
# Description:
# Using PsychHID to register keyboard events in PsychoPy

from psychopy.hardware import keyboard
from psychopy import core, visual

win = visual.Window((200, 200))

# Create a keyboard object
kb = keyboard.Keyboard()

# We define a function to print out the key presses
def waitKey():
    ''' a function to detect a single keypress'''

    got_key = False
    while not got_key:
        keys = kb.getKeys()
        if keys:
            for key in keys:
                print(key.name, key.duration, key.rt, key.tDown)
            got_key = True

# A simple response time task
# Change the window color following each key press
for i in range(10):
    win.color = (i % 2 * 1.0, -1, -1)
    win.flip()
    kb.clearEvents()
    kb.clock.reset()  # reset the clock
    waitKey()

# Close the window and quit PsychoPy
win.close()
core.quit()
```

In the example script, we first initialize a PsychHID Keyboard, which takes the following parameters.

```
keyboard.Keyboard(device=-1,
                  bufferSize=10000,
                  waitForStart=False,
                  clock=None)
```

The "device" parameter only applies to Linux and macOS, and it allows users to select from multiple keyboard devices, and then call *start()* and *stop()* to switch input from different keyboard devices. When set to −1, PsychoPy will use the default keyboard. The *bufferSize* parameter specifies how many keypresses to store in the buffer before it starts to drop the earliest keys. You can also provide a "clock" parameter to specify which *core.Clock*() object to use,[3] without which PsychoPy will use the global clock. The "waitForStart" parameter is usually set to *False*, as it is generally unnecessary to start/stop polling the keyboard manually.

To monitor keyboard events, in a *while* loop, we call *getKeys*() to retrieve a list of cached keypress events, with which the key name, duration (from keypress to key release), response time (from the most recent clock reset), and the timestamp of keypress can be extracted.

Trial Control

An experimental task usually involves repeated measurement of behaviors, and each measurement is known as a "trial." A trial in most eye-tracking experiments typically consists of multiple events. For example, the trial may start with the presentation of a fixation cross, followed by an image, which remains on the screen until a keyboard response. Implementing any experiment with multiple trials raises several important issues. How to present the trials in a randomized order? Which trials are grouped in blocks? Should a trial be recycled and presented to the subject again at a later time if an error occurs? All these questions need to be dealt with when considering the control of experimental trials. Here, I will present a very simple (yet flexible) way of controlling trial presentation using Python lists.

Before getting to the details, it is worth emphasizing that putting trials in a big for loop, which contains another for loop, is generally a terrible idea (I have seen too many MATLAB scripts like this). It makes the code difficult to read, even to the author himself/herself. Instead, it is usually preferable to define a "trial" function to handle the sequence of events that would occur in a single trial. Of course, this trial

[3] PsychoPy allows users to instantiate multiple clocks to keep track of time for different purposes, e.g., a global clock to keep track of the amount of time elapsed since task onset and another clock to calculate response time on each trial.

function should change the behavior of an experimental trial based on pre-specified parameters. Then, trial control becomes a simple matter of specifying the parameters of all trials in an iterable structure (e.g., a list) or a spreadsheet that the experimental script can read in at the beginning. The following pseudo-code illustrates this simple idea.

```
# Define a trial function
def run_trial(par1, par2, par3):
    '''present the trial events based on the parameters'''

    do_something here
    return trial_result

# Specify the parameters of all unique trials in a list
trials = [[t1_par1, t1_par2, t1_par3],
          [t2_par1, t2_par2, t2_par3],
          ...
          [tn_par1, tn_par2, tn_par3]]

# Iterate over all the trials
for trial_pars in trials:
    par1, par2, par3 = trial_pars
    run_trial(par1, par2, par3)
```

Assume that we would like to manipulate three parameters in each experimental trial (par1, par2, par3). In an actual experiment, these parameters may code features such as the size, color, or location of the visual stimuli. For a single trial, the values for these three parameters are stored in a list or a tuple, e.g., [t1_par1, t1_par2, t1_par3] for trial 1. We then put these lists into another list, as we did in the second step in the pseudo-code above. Then, we loop over the list and pass the relevant trial parameters to the trial function for each iteration of the loop.

If we need each of the unique trials to repeat 24 times, we can simply multiply the unique trial list by 24, e.g., *trial2Test = trials[:] *24*. If the trials are to be presented in a random order, we could call *random.shuffle*() from the *random* module, e.g., *random.shuffle(trial2Test)*.

The *run_trial* function in the above pseudo-code does not need to return any value, but a return value can be useful if you need to recycle error trials. For instance, the example script below allows the participant to press the right arrow key to pass a trial and to press the left arrow key to recycle a trial. If you press the right arrow key, the *run_trial()* function returns *False*; if you press the left arrow key, the *run_trial()* function returns True. In a while loop, we first randomize the trial list, and then we grab the last trial and *pop* it out of the *trial_list*. If the return value of this trial (*should_recycle*) is *True*, we add this trial back to the *trial_list*. The while loop will terminate if all trials have been completed successfully.

Figure 2.8 shows a screenshot of the following script in action. The trial currently being tested is "t3." If we keep pressing the left arrow key to recycle this trial, "t3" will remain in the *trial_list* and present itself again in a later iteration.

```
#!/usr/bin/env python3
#
# Filename: trial_recycle.py
# Author: Zhiguo Wang
# Date: 2/6/2021
#
# Description:
# This demo shows how to recycle a trial

import random
from psychopy import visual, event, core

win = visual.Window(size=(600, 400), units='pix')


def run_trial(trial_id):
    """a simple function to run a single trial"""

    # Show some info on the screen
    task_instruction = f'This is Trial: {trial_id}\n\n' + \
                       'RIGHT--> Next trial\n' + \
                       'LEFT--> Recycle current trial'
    msg = visual.TextStim(win, text=task_instruction)
    msg.draw()
    win.flip()

    # wait for a response
    key = event.waitKeys(keyList=['left', 'right'])

    # clear the window
    win.clearBuffer()
    win.flip()
    core.wait(0.5)

    if 'right' in key:
        return False
    if 'left' in key:
        return True

trial_list = ['t1', 't2', 't3', 't4', 't5']


# Recycle trials with a while-loop
while len(trial_list) > 0:
```

```
    # randomize the trial list
    random.shuffle(trial_list)

    # grab the last trial and pop it out the trial_list
    trial_to_test = trial_list.pop()

    # run a single trial
    should_recycle = run_trial(trial_to_test)

    # add the trial back to the trial_list if the
    # return value is True (i.e., need to recycle)
    if  should_recycle:
        trial_list.append(trial_to_test)

    # show what trials are left in the trial list
    print(f'The trial you just completed is: {trial_to_test}')
    print(f'Trials left in the list: {trial_list}')

# Close the window and quit PsychoPy
win.close()
core.quit()
```

The trial control method illustrated here is useful if you have a limited number of unique experimental conditions, like the Simon effect script presented in the following section. You may use a text file of comma-separated values to store the trial parameters. Then, you can read in the text file line by line to construct a list of trial parameters in your experimental script. You may also use the trial control routines (ExperimentHandler) provided by PsychoPy; detailed information can be found in the *data* section of the PsychoPy documentation.

A Real Example: Simon Effect

The previous sections have covered all the programming tools needed to build a simple experimental task. This section will go through a short script that implements the famous Simon effect. In this task, the participant responds to the color of a disk that could appear on either the left or right side of the screen. The color of the disk dictates which button (hand)—left or right—is the correct response, e.g., left button for a red disk and right button for a blue disk. The location of the disk is irrelevant. If a red disk appears on the left side of the screen, that will give us a *congruent* trial, as the disk location (left) is congruent with the correct response (left

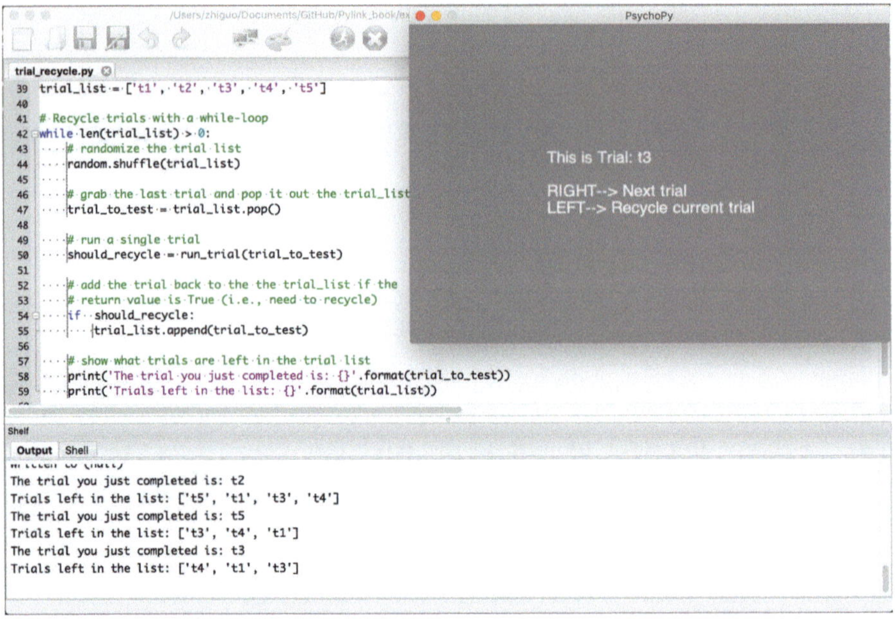

Fig. 2.8 An example script for trial recycling. In the while loop, we randomly select a trial from a trial list; it is then removed from the list if the return value (should_recyle) is False

button for red disks). However, if a blue disk appears on the left side, the trial is *incongruent* as the disk location (left) is not the same as the correct response (right button for blue disks). The "Simon effect" refers to the very robust finding that incongruent trials incur a significant response cost compared to congruent trials. In other words, reaction times are longer if the target requires a response from a hand on the opposite side, even though the target location is irrelevant. This is, of course, an oversimplified overview of the Simon effect; interested readers will find the review by Lu and Proctor (1995) helpful (Fig. 2.9).

The script for this task is listed below. The entire task can be implemented with less than 100 lines of code, including empty lines and comments.

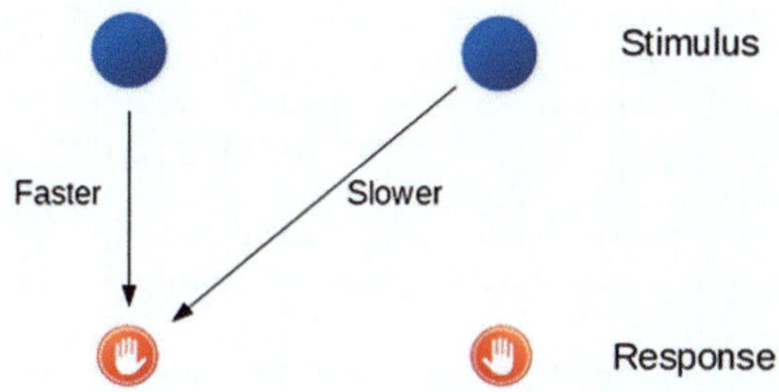

Fig. 2.9 *An illustration of the Simon task. The subject responds to the color of the disk; the response time is longer when the stimulus appears on the opposite side (relative to the responding hand)*

```
#!/usr/bin/env python3
#
# Filename: simon_effect.py
# Author: Zhiguo Wang
# Date: 2/6/2021
#
# Description:
# Measuring the Simon effect in PsychoPy

import random
from psychopy import visual, core, event, gui

# Open a window and prepare the stimuli
win = visual.Window((1280, 800), units='pix', fullscr=False, color='black')
text_msg = visual.TextStim(win, text='message')
tar_stim = visual.GratingStim(win, tex='None', mask='circle', size=60.0)

# Possible target position
tar_pos = {'left': (-200, 0), 'right': (200, 0)}

# A list of all possible trial parameter combinations
trials = [
    ['left', 'red', 'z', 'congruent'],
    ['left', 'blue', 'slash', 'incongruent'],
    ['right', 'red', 'z', 'incongruent'],
    ['right', 'blue', 'slash', 'congruent']
    ]

def run_trial(trial_pars, data_file, participant):
    """ Run a single trial.

    trial_pars -- target position, color, and correct key, e.g.,
                  ['left', 'red', 'z', 'cong']
```

```
    data_file -- a file to save trial data
    participant -- information about the participant in a dictionary,
                  {'id':1, 'name':zw}"""

    # Unpacking the parameter list
    pos, color, cor_key, congruency = trial_pars

    # Set target position and color
    tar_stim.pos = tar_pos[pos]
    tar_stim.color = color

    # Present a fixation cross for 750 ms
    text_msg.text = '+'
    text_msg.draw()
    win.flip()
    core.wait(0.750)

    # Present the target and wait for a key response
    tar_stim.draw()
    win.flip()
    t_tar_onset = core.getTime()
    tar_resp = event.waitKeys(1500, ['z', 'slash'], timeStamped=True)

    # write data to file
    trial_data = list(participant.values()) + \
        trial_pars + [t_tar_onset] + \
        list(tar_resp[0])
    trial_data = map(str, trial_data)  # convert list items to string
    data_file.write(','.join(trial_data) + '\n')

    # clear the screen and set an ITI of 500 ms
    win.color = 'black'
    win.flip()
    core.wait(0.500)

# ------ Real experiment starts here -------


# Get participant info with a dialog
participant = {'Participant ID': 0, 'Participant Initials': 'zw'}
dlg = gui.DlgFromDict(participant, title='Enter participant info here')
```

```
# Open a data file with write permission
d_file = open(participant['Participant Initials']+'.csv', 'w')

# Show task instructions
text_msg.text = 'Press Z to RED\nPress / to BLUE\n\nPress <SPACE> to start'
text_msg.draw()
win.flip()
event.waitKeys(keyList=['space'])

# Randomly shuffle the trial list and iterate over all of them
random.seed = 1000  # Set a random seed
trial_list = trials[:]*2
random.shuffle(trial_list)
for pars in trial_list:
    run_trial(pars, d_file, participant)

# Close the data file
d_file.close()

# Close the window and quit PsychoPy
win.close()
core.quit()
```

The script begins with the importing of necessary libraries and modules. Note that we also import Python's random module to handle trial randomization in addition to the frequently used PsychoPy modules (*visual*, *event*, and *core*).

```
import random
from psychopy import visual, core, event, gui
```

As usual, we need first to open a window; then, we create the target stimulus in memory with *visual.GratingStim*(). Note that setting the "tex" parameter to *None* and the "mask" parameter to "circle" will give us a *visual.GratingStim*() with maximum contrast (i.e., a filled disk).

```
# Open a window and prepare the stimuli
win = visual.Window((1024, 768), units='pix', fullscr=False, color='black')

tar_stim = visual.GratingStim(win, tex='None', mask='circle', size=60.0)
```

For this simple task, the parameters of a trial include target position, target color, correct response key, and congruency. Assume that the target appears on the left side of the screen; if the required response key is 'z', we will have a congruent trial; if the required response key is '/' (slash), we will have an incongruent trial. The participant must press 'z' to red targets and '/' to blue targets. Consider all possible values for each of these parameters, and we construct a list of unique trials.

```
# A list of all possible trial parameter combinations
trials = [
    ['left', 'red', 'z', 'congruent'],
    ['left', 'blue', 'slash', 'incongruent'],
    ['right', 'red', 'z', 'incongruent'],
    ['right', 'blue', 'slash', 'congruent']
    ]
```

Here the parameters of each unique trial are stored in a list, e.g., *['left', 'red', 'z', 'congruent']*; these lists are then stored in another list (*trials*). Note that, in the trial parameter list, '*left*' and '*right*' are used to specify the target position. The actual target position in screen pixel coordinates is stored in a dictionary called *tar_pos*. So, *tar_pos['left']* will give us (–200, 0), and *tar_pos['right]* will give us (200, 0).

```
# Possible target position
tar_pos = {'left': (-200, 0), 'right': (200, 0)}
```

The most crucial bit of this script is, of course, the *run_trial()* function. This function takes three arguments: a list that stores the trial parameters (trial_*pars*), a file to log the data collected from each trial (*data_file*), and a dictionary storing participant information (*participant*).

```
def run_trial(trial_pars, data_file, participant):
    """ Run a single trial.

    trial_pars -- target position, color, and correct key, e.g.,
                  ['left', 'red', 'z', 'cong']
    data_file -- a file to save trial data
    participant -- information about the participant in a dictionary,
                  {'id':1, 'name':zw}"""
```

At the beginning of each trial, we unpack the trial parameters (*trial_pars*) to update the stimulus position and color, etc. Then, we present the fixation cross and the target, and wait for the subject to respond.

```
# Unpacking the parameter list
pos, color, cor_key, congruency = trial_pars

# Set target position and color
tar_stim.pos = tar_pos[pos]
tar_stim.color = color
```

The code at the end of the trial function may look rather complex at first but is just a sequence of steps that convert all the trial data into a format that can be written to a text file. First, we concatenate the participant information, the trial parameters, and the response data into a long list. An explicit line break ("\") is used to allow a long statement to span over multiple lines.

```
# write data to file
trial_data = list(participant.values()) + \
    trial_pars + [t_tar_onset] + \
    list(tar_resp[0])
```

We then convert all items in the list into strings with the *map* function, which applies the *str* function to all items in *trial_data*.

```
trial_data = map(str, trial_data)  # convert to string
```

Then, we use the string method *join()* to connect the items in the list with commas. The write() method is used to save the resulting string to a text file (*data_file*).

```
data_file.write(','.join(trial_data) + '\n')
```

Although this data logging method may appear somewhat low-tech, it is highly reliable. We write data to a plain text file at the end of each trial; so, even if a task crashes in the middle of a testing session, data from all the trials run so far will already have been saved. I have never lost any data with this simple data logging method. The resulting .csv file can be loaded into Excel or R for further examination (***see*** Fig. 2.10).

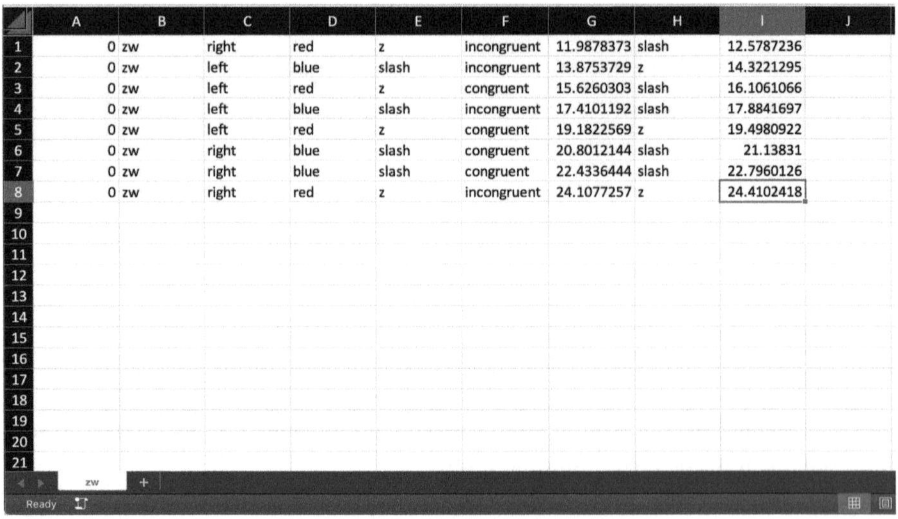

	A	B	C	D	E	F	G	H	I	J
1	0	zw	right	red	z	incongruent	11.9878373	slash	12.5787236	
2	0	zw	left	blue	slash	incongruent	13.8753729	z	14.3221295	
3	0	zw	left	red	z	congruent	15.6260303	slash	16.1061066	
4	0	zw	left	blue	slash	incongruent	17.4101192	slash	17.8841697	
5	0	zw	left	red	z	congruent	19.1822569	z	19.4980922	
6	0	zw	right	blue	slash	congruent	20.8012144	slash	21.13831	
7	0	zw	right	blue	slash	congruent	22.4336444	slash	22.7960126	
8	0	zw	right	red	z	incongruent	24.1077257	z	24.4102418	

Fig. 2.10 The output from a testing session of eight trials

The above code is just preparatory work. The real task starts with a simple dialog created with the *gui* module of PsychoPy. In the script, we use a dictionary to construct the dialog and store the participant information entered by the experimenter.

```
participant = {'Participant ID': 0, 'Participant Initials': 'zw'}
dlg = gui.DlgFromDict(participant, title='Enter participant info here')
```

We then open a text file, so it could be used to store experimental data later on.

```
# Open a data file with write permission
d_file = open(participant['Participant Initials']+'.csv', 'w')
```

The present task has four unique trials, and the parameters of all unique trials are stored in a list (*trials*). The next step is to construct a list of trials (*trial_list*) that we would like to test. It is essential to understand that computers can only do pseudo-randomization. With a fixed random seed, though the trials are randomized, all participants will complete the trials in the same sequence. So, using a fixed random seed may not be preferable in all tasks.

```
random.seed = 1000  # Set a random seed
trial_list = trials[:]*2
random.shuffle(trial_list)
```

To test all experimental trials, we need the help of a *for* loop. On each iteration of the *for* loop, we pass the parameters of each trial (*pars*), the data file we just opened (*d_file*), and the participant information we collected from the dialog (*participant*) to the *run_trial*() function. The *run_trial*() function will present the stimuli, register keyboard responses, and write collected data to file. The *for* loop terminates when all trials have been completed.

```
for pars in trial_list:
    run_trial(pars, d_file, participant)
```

After all trials have been presented, the script will close the data file, close the window, and quit PsychoPy.

```
# Close the data file
d_file.close()

# Close the window and quit PsychoPy
win.close()
core.quit()
```

The features of PsychoPy and Python covered in this chapter should be sufficient for implementing many experimental tasks. The example code that comes with PsychoPy covers almost all aspects of PsychoPy and is a perfect place to get started or find out more about PsychoPy and its features. The online reference manual of PsychoPy is also a great source of information that I highly recommend.

Chapter 3
Building Experiments with Pygame

Contents

Pygame is a Python wrapper for the popular SDL multimedia library. [1] Pygame is suitable for tasks like creating 2D graphics and monitoring keyboard, mouse, and gamepad events. It may appear redundant to have a chapter on Pygame since we have already covered PsychoPy, which offers good timing precision and a rich feature set that is easy to use. However, while PsychoPy depends on about 60 other Python modules, Pygame requires no additional dependencies. It is a lightweight solution appealing to some researchers. Exploring Pygame will also allow us to consolidate the basic Python skills we have discussed in Chaps. 1 and 2.

In the first chapter of this book, I mentioned a few Python modules designed for building psychological experiments, e.g., PsychoPy. These modules have been optimized for precise timing, though some libraries perform better than others (for a comparison, see Bridges et al., 2020). If you use Pygame to script a timing-critical experimental task, do not forget to verify timing precision and reliability.

[1] For an overview of Pygame, see https://www.pygame.org/docs/tut/PygameIntro.html.

Z. Wang, *Eye-Tracking with Python and Pylink*,
https://doi.org/10.1007/978-3-030-82635-2_3

Installing Pygame

Pygame can be easily installed with "pip" (see Chap. 1 for instructions). Pygame released a major update (version 2.0) while this book was being written. *pip,* by default, will install version 2.0, but you can install the legacy version (1.9.6) if needed. The example scripts presented in this chapter should work fine with both Pygame versions 1.9.6 and 2.0.

To test if you have successfully installed Pygame, start a Python shell, and enter the following command to see if you can import Pygame without issue.

```
import pygame
```

Pygame offers multiple functional modules, including *display*, *draw*, *font*, *image*, *key*, *mixer*, *mouse*, *time*, etc. As noted in Chap. 1, an in-depth discussion of these modules can be found in *Beginning Game Development with Python and Pygame: From Novice to Professional* by Will McGugan. In this chapter, we will introduce the most frequently used modules through examples. Before we delve in, there are a few things to clarify.

Note that most Pygame modules require "initialization" before you can use their various functions and features. For instance, to play audio files and present texts on-screen, we need to initialize the *mixer* and *font* modules.

```
pygame.mixer.init()
pygame.font.init()
```

One common practice is to call *pygame.init*() to initialize all modules at the beginning of the script. The downside of this approach is that Pygame may complain or crash if some modules are not available (e.g., loading the *mixer* module without a proper sound driver).

The Pygame library includes lots of constants; for instance, *pygame.K_z* represents the ASCII code for the letter "z" (i.e., 122). It can be cumbersome to refer to these constants with the "pygame." prefix. To avoid this, you may import all the "local" constants of the Pygame library at the beginning of the scripts.

```
from pygame.locals import *
```

Display

As with PsychoPy, the first step in creating an experimental task is to open a window. A window, by its very nature, is a surface, which represents an image. Opening a window is much like getting a whiteboard on which you draw various geometric shapes or attach an image onto it. In Pygame, we use the *set_mode*() function from the display module to open a window.

```
pygame.display.set_mode(size=(0, 0), flags=0, depth=0, display=0)
```

pygame.display.set_mode() takes several parameters. The size parameter requires a tuple specifying the size of the window in pixels. If a window is open in full-screen mode, but the required window size is not one of the supported screen resolutions, Pygame will use the closet match. The depth parameter is usually not set so that Pygame will default to the best and fastest color depth for the system. The parameter that we do need to specify is the flags, which determine what type of window to initialize. We can combine multiple flags through a bitwise OR operation by separating the options with the pipe ("|") character. For better timing performance, in most experimental scripts, we should supply the FULLSCREEN, DOUBLEBUF, and HWSURFACE flags to tap into the full potential of modern graphics cards (e.g., hardware acceleration).

```
pygame.display.set_mode((1440, 900), HWSURFACE|DOUBLEBUF|FULLSCREEN)
```

The display module also comes with a handy tool for retrieving all supported screen resolutions, i.e., *pygame.display.list_modes()*. Use this function only after you have called *pygame.init()* or *pygame.display.init()*.

```
>>> pygame.display.list_modes()

[(2560, 1600), (2048, 1280), (1650, 1050), (1440, 900), (1280, 800), (1152,
720), (1024, 768), (800, 600), (840, 524), (640, 480)]
```

Another frequently used function is *pygame.display.get_surface()*, which returns the currently active window. In the custom drawing function defined in the pseudo-code below, *pygame.display.get_surface()* is called to give us a reference to the active window so that we could draw on it.

```
pygame.display.set_mode((600, 400))

def my_draw_circle(size, where):
    ''' draw a circle '''

    surf = pygame.display.get_surface()
    pygame.draw.circle(surf, color, where, size)
    pygame.display.flip()
```

Most modern graphics cards support double buffering. With double buffering, the computer draws the graphics in the back buffer. To show the graphics on the screen, we need to call *pygame.display.flip()* to flip the back buffer to the front, as shown in the pseudo-code above.

```
pygame.display.flip()
```

If you set a display mode with the HWSURFACE and DOUBLEBUF flags, *pygame.display.flip()* will wait for a vertical retrace before it swaps the graphics buffers. So, it is possible to examine the monitor refresh intervals by recording the time interval between two consecutive calls of *pygame.display.flip*(). This trick is illustrated in the short script below.

In the script, we first import Pygame and its local constants, initialize the modules, get the native resolution supported by the monitor, and then open a full-screen window with double buffering and hardware acceleration. We then record the

```
#!/usr/bin/env python3
#
# Filename: display_demo.py
# Author: Zhiguo Wang
# Date: 2/11/2021
#
# Description:
# Open a window to assess monitor refresh consistency

import pygame
from pygame.locals import *

# Initialize Pygame & its modules
pygame.init()
```

```
# Get the native resolution supported by the monitor
scn_res = pygame.display.list_modes()[0]

# Open a window
win = pygame.display.set_mode(scn_res, DOUBLEBUF | HWSURFACE | FULLSCREEN)

# An empty list to store the monitor refresh intervals
intv = []

# Flip the video buffer, then grab the timestamp of the first retrace
pygame.display.flip()

# Get the timestamp of the 'previous' screen retrace
t_before_flip = pygame.time.get_ticks()

# Use a for-loop to flip the video buffer 200 times
for i in range(200):
    # Switching the window color between black and white
    if i % 2 == 0:
        win.fill((255, 255, 255))
    else:
        win.fill((0, 0, 0))
    # Flip the video buffer to show the screen
    pygame.display.flip()

    # Get the timestamp of the 'current' screen retrace
    t_after_flip = pygame.time.get_ticks()
    # Get the refresh interval
    flip_intv = t_after_flip - t_before_flip
    # Store the refresh interval to "intv"
    intv.append(flip_intv)
    # Reset the timestamp of the 'previous' retrace
    t_before_flip = t_after_flip

# Print out the max, min, and average refresh intervals
intv_max = max(intv)
intv_min = min(intv)
intv_avg = sum(intv)*1.0/len(intv)
print('Max: {}, Min: {}, Mean: {}'.format(intv_max, intv_min, intv_avg))

# Quit Pygame
pygame.quit()
```

timestamp of each call of *pygame.display.flip()*, calculate the screen refresh interval, and store it in a list (*intv*).

In the script, *pygame.time.get_ticks*() is used to retrieve the number of milliseconds that have elapsed since the call to *pygame.init*(). Before entering the *for* loop, this function is called to give us a starting timestamp (*t_before_flip*).

```
# Flip the video buffer, then grab the timestamp of the first retrace
pygame.display.flip()

# Get the timestamp of the 'previous' screen retrace
t_before_flip = pygame.time.get_ticks()
```

In the *for* loop, we get the time following each window flip (*t_after_flip*), calculate the screen refresh interval (*flip_intv*), and append the interval to the list (*intv*) that we initialized outside the *for* loop.

```
    # Get the timestamp of the 'current' screen retrace
    t_after_flip = pygame.time.get_ticks()
    # Get the refresh interval
    flip_intv = t_after_flip - t_before_flip
    # Store the refresh interval to "intv"
    intv.append(flip_intv)
    # Reset the timestamp of the 'previous' retrace
    t_before_flip = t_after_flip
```

Once the *for* loop is done, we print out the minimum, maximum, and average refresh interval. If everything works perfectly, you would expect a mean interval close to the duration of a retrace cycle. For instance, with a monitor capable of redrawing graphics at 60 Hz, the duration of a retrace cycle is 16.67 ms.

```
# Print out the max, min, and average refresh intervals
intv_max = max(intv)
intv_min = min(intv)
intv_avg = sum(intv)*1.0/len(intv)
print('Max: {}, Min: {}, Mean: {}'.format(intv_max, intv_min, intv_avg))
```

Running this script in IDLE will give your graphics card and monitor a quick test. The test results for my MacBook Pro, which has an integrated Intel graphics card and a 60-Hz built-in screen, were not great. The mean screen refresh interval was 19.58 ms, much longer than what we would expect for a 60-Hz screen

(16.67 ms). The results for my Windows PC, which has an Nvidia GeForce GTX graphics card and a 144-Hz BenQ monitor, were reasonably decent. The comparison here is about the graphics card and monitors, not the operating systems. We can also achieve reliable timing on Macs equipped with discrete graphics cards.

```
Max: 139, Min: 8, Mean: 19.58  # Results from my MacBook Pro
Max: 8, Min: 2, Mean: 6.89  # Results from my desktop PC
```

Events

To create an interactive application, it is necessary to understand how Pygame detects and handles various "events." For instance, when you click the "close" button of a window, Pygame will generate a QUIT event; when you press a key down, Pygame will generate a KEYDOWN event. There are several other types of events that Pygame natively supports, for instance, MOUSEMOTION (mouse motion) and JOYBUTTONDOWN (joystick button down).

Pygame places all events in a queue. The preferred event handling function is *pygame.event.get()*, which grabs all the queued events and then empties the queue. *pygame.event.get()* will return a list of events added to the queue since the last call of this function.

```
[<Event(12-Quit {})>]
[<Event(5-MouseButtonDown {'pos': (192, 124), 'button': 1, 'window':
None})>]
[<Event(2-KeyDown {'unicode': 'a', 'key': 97, 'mod': 0, 'scancode': 0,
'window': None})>]
[<Event(4-MouseMotion {'pos': (0, 185), 'rel': (0, 1), 'buttons': (0, 0,
0), 'window': None})>]
```

As shown in the shell output above, *pygame.event.get()* returns event(s) in a list. Each event contains a unique code; for instance, the event code for "MouseMotion" is 4. These event codes are Pygame constants; typing "pygame.MOUSEMOTION" in a Python shell will return the event code 4.

```
>>> pygame.MOUSEMOTION
4
```

An event also contains information other than the event code. In the MOUSEMOTION event shown above, the event properties that can be retrieved include '*pos*', '*rel*', and '*buttons*', which correspond to the current mouse position, (0, 185); the distance the mouse has moved since the previous MOUSEMOTION event, (0, 1); and the status of the mouse buttons, (0, 0, 0).

```
while True:
    ev_list = pygame.event.get()
    for ev in ev_list:
        if ev.type == MOUSEMOTION:
            print(ev.pos)
```

The above code snippet monitors and retrieves the mouse position. In a *while* loop, *pygame.event.get()* is first called to return a list of events that we can loop over to see if there is a MOUSEMOTION event. If so, we print out the mouse "pos" stored in this event. We can use a similar method to retrieve keypresses and other events. The following example script will print out the mouse position if the mouse moves, print out the button number if a mouse click occurs, print out the ASCII code of a key if pressed, and terminate the script if the "close" button on the window is clicked.

```
#!/usr/bin/env python3
#
# Filename: event_demo.py
# Author: Zhiguo Wang
# Date: 2/7/2021
#
# Description:
# A short script showing how to handle Pygame events

import sys
import pygame
from pygame.locals import *

# Initialize Pygame and open a window
pygame.init()
scn = pygame.display.set_mode((640, 480))

# Constantly polling for new events in a while-loop
while True:
    ev_list = pygame.event.get()
    for ev in ev_list:
        # Mouse motion
        if ev.type == MOUSEMOTION:
```

```
            print(ev.pos)

        # Mouse button down
        if ev.type == MOUSEBUTTONDOWN:
            print(ev.button)

        # Key down
        if ev.type == KEYDOWN:
            print(ev.key)

        # Quit Pygame if the "close window" button is pressed
        if ev.type == QUIT:
            pygame.quit()
            sys.exit()
```

Other Frequently Used Pygame Modules

The draw *Module*

For simple drawings (e.g., shapes and lines), use the *draw* module. Frequently used functions available in the *draw* module include *draw.polygon*(), *draw.circle*(), *draw.line*(), *draw.rect*(), *draw.ellipse*(), etc. The quality of geometric shapes created with this module is not great but should be sufficient for tasks requiring simple graphics, like the Posner cueing task presented in this chapter.

In the script below, we append the mouse position to a list (*points*) every time the mouse is clicked. When the list has more than three points, we draw a polygon and use filled circles to mark the vertices (Fig. 3.1).

```
#!/usr/bin/env python3
#
# Filename: draw_demo.py
# Author: Zhiguo Wang
# Date: 2/7/2021
#
# Description:
# A script illustrating the drawing functions in Pygame
```

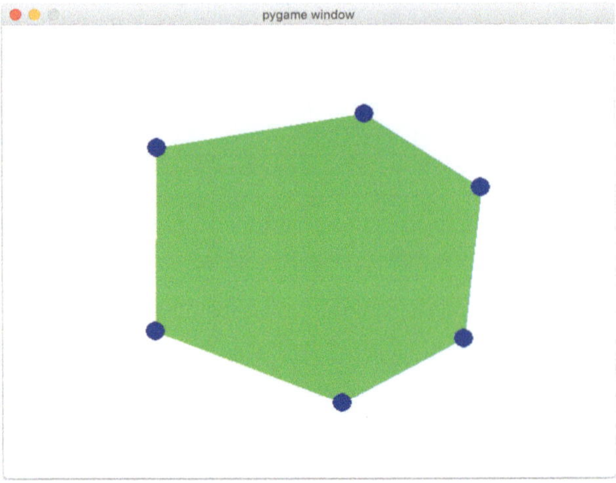

Fig. 3.1 Draw a polygon and show the vertices in Pygame

```
import pygame
import sys
from pygame.locals import *

# Initialize pygame and open a window
pygame.init()

# Open a window
scn = pygame.display.set_mode((640, 480))

# An empty list to store clicked screen positions
points = []

while True:
    # Poll Pygame events
    for ev in pygame.event.get():
        # Quit Pygame and Python if the "close window"
        # button is clicked
        if ev.type == QUIT:
            pygame.quit()
            sys.exit()

        # Append the current mouse position to the list when
        # a mouse button down event is detected
        if ev.type == MOUSEBUTTONDOWN:
            points.append(ev.pos)

    # Clear the screen
    scn.fill((255, 255, 255))
```

```
    # Draw a polygon after three mouse clicks
    if len(points) >= 3:
        pygame.draw.polygon(scn, (0, 255, 0), points)

    # Highlight the screen locations that have been clicked
    for point in points:
        pygame.draw.circle(scn, (0, 0, 255), point, 10)

    # Flip the video buffer to show the drawings
    pygame.display.flip()
```

Calling *pygame.quit()* will un-initialize all Pygame modules. Consequently, the drawing functions may crash in the *while* loop. To gracefully terminate the script, we call *sys.exit()* to quit the Python interpreter altogether when the "Close" window button is clicked.

The font *Module*

With Pygame, we need to initialize a font object to render text on the screen. To create a font object, call *pygame.font.SysFont()* to use a system font or *pygame.font.Font()* to use a custom TrueType[2] font file instead. For cross-platform applications, you may prefer *pygame.font.Font()* as the font you choose to use may not be available on all platforms, and the name of a font may differ across platforms. You can also specify a list of system fonts, so Pygame will exhaust these fonts before reverting to the "default" font. The "default" font can be retrieved by *pygame.font.get_default_font()*. To list all fonts available on your OS, call *pygame.font.get_fonts()*.

```
>>> pygame.font.get_default_font()
'freesansbold.ttf'
>>> pygame.font.get_fonts()
['cochin', 'iowanoldstyle', 'avenir', 'nanumgothic',
'bitstreamverasansmono',
...
'bodoniornaments', 'weibeitc', 'libiansc', 'plantagenetcherokee']
```

The short script below shows how to render text in Pygame. To present text, we need first to initialize a font object (*fnt*), which controls the appearance of the text. The script illustrates two important methods available to a font object, *size*() and *render*(). The *render*() method will convert texts into a surface, which can be drawn

[2] TrueType is a font standard developed by Apple and Microsoft in the late 1980s. It is the most common font format on macOS and Windows.

onto a Pygame window with *blit*(). The *size*() method returns the dimension of such a text surface when we pass text ("Hello, World!") to it.

```
#!/usr/bin/env python3
#
# Filename: text_demo.py
# Author: Zhiguo Wang
# Date: 2/7/2021
#
# Description:
# Text rendering example in Pygame

import pygame, sys


# Initialize Pygame
pygame.init()

# Open a window
win = pygame.display.set_mode((300,200))

# Create a font object and enable 'underline'
fnt = pygame.font.SysFont('arial', 32, bold=True, italic=True)
fnt.set_underline(True)

# Using the size() method to estimates the width and height of
# the rendered text surface
demo_text = 'Hello, World!'
w, h = fnt.size(demo_text)

# Render the text to get a text surface
win.fill((0, 0, 0))
text_surf = fnt.render(demo_text, True, (255,0,0))

# Show(blit) the text surface at the window center
win.blit(text_surf, (150-w/2,100-h/2))
pygame.display.flip()

# Show the text until a key is pressed
while True:
    for ev in pygame.event.get():
        if ev.type == pygame.KEYUP:
            pygame.quit()
            sys.exit()
```

For a font object, underlining and bolding can be enabled or disabled with the *set_underline()* and *set_bold()* methods. The method *set_italic()*, however, only allows "fake rendering of italic text," which skews the font in a way that "doesn't look good on many font types" (see the Pygame documentation). The *font* module does not support text wrapping, so there is no easy solution for presenting multiline texts. Nevertheless, you can use *size()* to figure out the dimension of each word in the texts, to help determine which word will reach the edge of the window and when to render a new line.

The image *Module*

The *image* module allows users to easily manipulate images in Pygame. Pygame supports frequently used formats, such as BMP, PNG, and JPEG. The Pygame function for loading images is *pygame.image.load()*, which will return a surface that can be drawn onto a Pygame window with *blit()*.

```
win = pygame.display.set_mode((800, 600))  # open a window
img = pygame.image.load('test.png') # load an image
win.blit(img, (0,0)) # blit the image to the top-left corner of the window
```

A Real Example: Posner Cueing Task

As noted, Pygame is perfect for simple experimental tasks, such as the classic Posner cueing task (Posner 1980). In the example below, I will use the task procedure of a recent publication by Fu et al. (2019), where we examined if dyslexic children are slower to shift spatial attention when compared to healthy controls. The design of the task is relatively straightforward. There are always three boxes (placeholders) on the screen, and the central one contains a fixation cross. After some time, one of the two peripheral boxes is cued by making its border thicker for a brief time. Then a circular target appears, either in the cued box or in the other box. Participants are required to make a speeded response to the target. The interstimulus interval (ISI) between the cue and the target is manipulated to reveal the time course of attention orienting.

Figure 3.2 illustrates the sequence of events that occur in a single trial. The initial fixation screen is presented for 1000 ms; then, one of the peripheral placeholders flashes (cue) briefly. After an ISI of 0, 100, 300, or 700 ms (randomized within blocks of trials), the target appears in either the cued peripheral box or the uncued

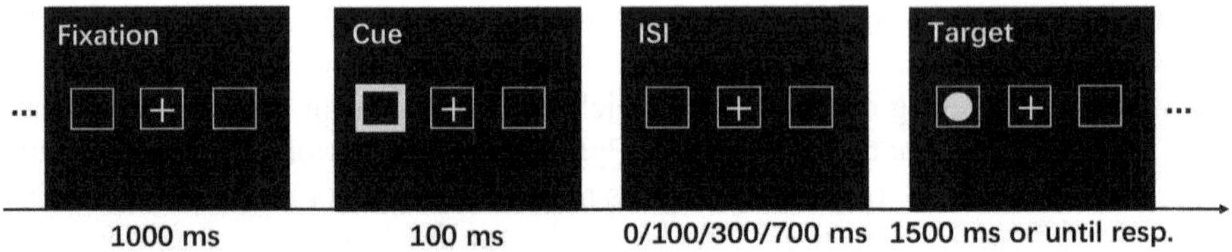

Fig. 3.2 The Posner cueing task used in Fu et al. (2019)

one. This experiment also monitored eye movements, but the relevant code is not included here for simplicity .

```
#!/usr/bin/env python3
#
# Filename: posner_cueing.py
# Author: Zhiguo Wang
# Date: 2/7/2021
#
# Description:
# A Posner cueing task implemented in Pygame

import random
import pygame
import sys
from pygame import display, draw, Rect, time, event, key, mouse
from pygame.locals import *

# Set a few constants
sz = 90  # size of the placeholder
colors = {'gray': (128, 128, 128),
          'white': (255, 255, 255),
          'black': (0, 0, 0)}
pos = {'left': (212, 384),
       'center': (512, 384),
       'right': (812, 384)}

# List of all unique trials, [cue_pos, tar_pos, isi, cueing, cor_key]
trials = []
for cue_pos in ['left', 'right']:
    for tar_pos in ['left', 'right']:
        for isi in [0, 100, 300, 700]:
            if cue_pos == tar_pos:
                cueing = 'cued'
            else:
                cueing = 'uncued'
            if tar_pos == 'left':
```

```
                cor_key = 'z'
            else:
                cor_key = '/'
            trials.append([cue_pos, tar_pos, isi, cueing, cor_key])

def draw_frame(frame, trial_pars):
    ''' Draw the possible screens.

    frame -- which frame to draw, e.g., 'fix', 'cue', 'target'
    trial_pars--parameters, [cue_pos, tar_pos, isti, cueing, cor_key]'''

    # Unpack the trial parameters
    cue_pos, tar_pos, isi, cueing, cor_key = trial_pars

    # Clear the screen and fill it with black
    win.fill(colors['black'])

    # The place holders are visible on all screens
    # Here, 'pos' is a dictionary;
    # we retrieve both the key and value pairs in a for-loop
    for key, (x, y) in pos.items():
        # Draw the place holder
        draw.rect(win, colors['gray'], Rect(x - sz/2, y - sz/2, sz, sz),1)

        # The fixation cross is visible on all screens
        if key == 'center':
            draw.line(win, colors['gray'], (x - 20, y), (x + 20, y), 3)
            draw.line(win, colors['gray'], (x, y - 20), (x, y + 20), 3)

    # Draw the fixation screen-- three placeholders with a cross
    if frame == 'fix':
        pass

    # Draw the cue (a bright box--a Rect)
    if frame == 'cue':
        c_x, c_y = pos[cue_pos]  # coordinates of the cue
        draw.rect(win, colors['white'], Rect(c_x - sz/2, c_y - sz/2,
                                             sz, sz), 5)

    # Draw the target (a filled white disk)
    if frame == 'target':
        draw.circle(win, colors['white'], pos[tar_pos], 20)

    display.flip()
```

```
def run_trial(trial_pars, subj_info, data_file):
    ''' Run a single trial.

    trial_pars -- a list specifying trial parameters,
                  [cue_pos, tar_pos, isi, cueing, cor_key]
    subj_info -- info about the subject [id, name, age]
    data_file -- an open file to save the trial data.'''

    # Show the fixation then wait for 1000 ms
    draw_frame('fix', trial_pars)
    time.wait(1000)

    # Show the cue for 100 ms
    draw_frame('cue', trial_pars)
    time.wait(100)

    # Inter-stimulus interval (ISI)
    draw_frame('fix', trial_pars)
    time.wait(trial_pars[2])

    # Show the target and register a keypress response
    draw_frame('target', trial_pars)
    tar_onset = time.get_ticks()
    tar_resp = -32768  # response time
    resp_key = -32768  # key pressed

    # Check for key presses
    time_out = False
    got_key = False
    event.clear()  # clear bufferred events
    while not (time_out or got_key):
        # Cehck for time out (1500 ms)
        if time.get_ticks() - tar_onset > 1500:
            time_out = True

        # Check if any key has been pressed
        for ev in event.get():
            if ev.type == KEYDOWN:
                if ev.key in [K_z, K_SLASH]:
                    tar_resp = time.get_ticks()
                    resp_key = key.name(ev.key)
                    got_key = True
```

```
    # write data to file
    trial_data = subj_info + trial_pars + [tar_onset, tar_resp, resp_key]
    trial_data = map(str, trial_data)
    data_file.write(','.join(trial_data) + '\n')

    # ITI (inter-trial_interval)
    draw_frame('fix', trial_pars)
    time.wait(1500)

# -- Real experiment starts from here --
# Get subject info from the Python shell
subj_id = input('Subject ID (e.g., 01): ')
subj_age = input('Subject Age: ')
subj_info = [subj_id, subj_age]

# Open a CSV file to store the data
d_file = open('d_{}.csv'.format(subj_info[0]), 'w')

# Open a window, add the FULLSCREEN flag for precise timing
win = display.set_mode((1024, 768), HWSURFACE | DOUBLEBUF)
# Hide the mouse cursor
mouse.set_visible(False)

# Randomly shuffle the trial list and test them one by one
test_trials = trials[:]*1  # how many trials to test
random.shuffle(test_trials)  # randomize
for pars in test_trials:
    run_trial(pars, subj_info, d_file)

# Close the data files
d_file.close()

# Quit Pygame
pygame.quit()
sys.exit()
```

The structure of the script is similar to that of the PsychoPy example presented in the previous chapter. The script starts with importing the necessary Pygame modules and defining constants, e.g., the size and positions of the placeholders, and the colors used in the task (gray, white, and black).

```
# Set a few constants
sz = 90  # size of the placeholder
colors = {'gray': (128, 128, 128),
          'white': (255, 255, 255),
          'black': (0, 0, 0)}
pos = {'left': (212, 384),
       'center': (512, 384),
       'right': (812, 384)}
```

The simple trial control approach detailed in the previous chapter is also used in this task. For a given trial, we need to manipulate the cue and target location, the cue-target ISI, and of course, we need to know which key is the correct one to press. Putting these variables in a Python list should give you something like *[cue_pos, tar_pos, ITI, cor_key]*. With nested *for* loops, the following lines of code put all possible unique trials in a list. Note that I added another variable *cueing,* which can be "cued" (target in the cued box) or "uncued" (target in the uncued box), to the list. This variable will ease the task of deriving a cueing effect (cued-uncued) during data analysis.

```
# List of all unique trials, [cue_pos, tar_pos, isi, cueing, cor_key]
trials = []
for cue_pos in ['left', 'right']:
    for tar_pos in ['left', 'right']:
        for isi in [0, 100, 300, 700]:
            if cue_pos == tar_pos:
                cueing = 'cued'
            else:
                cueing = 'uncued'
            if tar_pos == 'left':
                cor_key = 'z'
            else:
                cor_key = '/'
            trials.append([cue_pos, tar_pos, isi, cueing, cor_key])
```

As shown in Fig. 3.2, the task involves the drawing of multiple screens, i.e., a fixation screen that contains a fixation cross and three placeholders (empty boxes), a cue screen that includes the same elements but brightening one of the placeholders, and a target screen with the fixation cross, the placeholders, and the target (a bright disk). Here we define a simple function that we can use to draw all the screens. This function will take two arguments, *frame* and *trial_pars*. The argument *frame* is a label for the three types of the screen; it could be 'fix', 'cue', or 'target'. The argument *trial_pars* is a list that controls the behavior of each trial, e.g., where the target would appear.

```
def draw_frame(frame, trial_pars):
    ''' Draw the possible screens.

    frame -- which frame to draw, e.g., 'fix','cue', 'target'
    trial_pars -- parameters, [cue_pos, tar_pos, isi, cueing, cor_key]'''
    ......
```

One notable departure from the PsychoPy example is that we need to write a short routine to monitor keyboard events. After the target appears on the screen, we get the current timestamp and clear the event buffer. Then, we use a *while* loop to detect a keypress or until 1500 ms have elapsed. In PsychoPy, all we needed to do was call the *waitKeys()* function from the *event* or the *hardware* module.

```
    # Show the target and register a keypress response
    draw_frame('target', trial_pars)
    tar_onset = time.get_ticks()
    tar_resp = -32768  # response time
    resp_key = -32768  # key pressed

    # Check for key presses
    time_out = False
    got_key = False
    event.clear()  # clear bufferred events
    while not (time_out or got_key):
        # Check for time out (1500 ms)
        if time.get_ticks() - tar_onset > 1500:
            time_out = True

        # Check if any key has been pressed
        for ev in event.get():
            if ev.type == KEYDOWN:
                if ev.key in [K_z, K_SLASH]:
                    tar_resp = time.get_ticks()
                    resp_key = key.name(ev.key)
                    got_key = True
```

This chapter introduced the most frequently used modules of Pygame and presented a complete example illustrating the various Pygame functions that can be used to implement an experimental task. The Pygame library is designed for computer programmers, not for experimental psychologists. As the example presented here demonstrates, scripts in Pygame tend to be longer and less straightforward than those in PsychoPy. Nevertheless, Pygame is still a useful multimedia library that can be used to implement a wide range of experimental tasks.

Chapter 4
Getting to Know Pylink

Contents

A Brief Introduction to Eye Tracking

Put your thumb on this page, keep looking at the thumbnail, and you will see that words on either side of the thumb are somewhat challenging to read unless you stop focusing on the thumbnail. Why? Because most light-sensitive cells (cones) concentrate in a tiny area on the retina, known as the fovea, and visual acuity reduces

Z. Wang, *Eye-Tracking with Python and Pylink*,
https://doi.org/10.1007/978-3-030-82635-2_4

exponentially as the distance from the fovea increases. The fovea corresponds to about 1–2 degrees of visual angle. One degree is about the size of a thumbnail at arms' length. To examine the objects in the visual field, the eyes need to quickly rotate to change the gaze direction around 3–4 times a second. These rapid eye movements are known as "saccades." Between saccades are periods during which the eyes remain relatively still, so the processing of visual details can take place. These periods of relative stability are known as "fixations."

Since the 1870s, researchers have developed a wide range of methods to record eye movements, including electromagnetic and optical techniques. Currently, most commercial eye-tracking devices have adopted a video-based solution, which has the advantages of being comfortable to use and delivering the high levels of accuracy and precision required for laboratory research.

The working principle of a video-based eye tracker is illustrated in Fig. 4.1. An infrared (IR) light source illuminates the eye area, and a camera images the eyes at high speed (e.g., 1000 Hz). The camera filters out the visible light to obtain a clear image of the eyes in the infrared spectrum. In addition to illuminating the eyes, the infrared light source creates a bright "glint" on the cornea, known as the "corneal reflection," or CR for short. When the eyes rotate, the position of the pupil changes relative to the CR. We use image analysis software to compute the vector between the centers of the pupil and CR (pupil minus CR) in the camera sensor coordinates. This vector changes when the eye rotates, but not when the head moves (within 1 cubic inch space; see Merchant et al. 1974). As a result, small head movements are tolerable, and it is not necessary to use a bite bar to keep the head perfectly still.

For the tracker to report the gaze position in world-centered coordinates, a calibration process is required to create a mapping function between the raw pupil-CR data in camera sensor coordinates and the spatial location the subject is currently fixating (e.g., in screen pixel coordinates). The calibration result is a complex regression model that allows the eye tracker to estimate where the participant is looking (i.e., gaze position).

SR Research EyeLink® eye trackers are used in a wide range of research settings. The use of the tracker typically involves launching an experimental script on a stimulus presentation PC (or Display PC, in EyeLink terminology), adjusting the camera for optimal tracking, calibrating the tracker, evaluating the calibration results, and then starting the task to record eye movement data. The operation of the

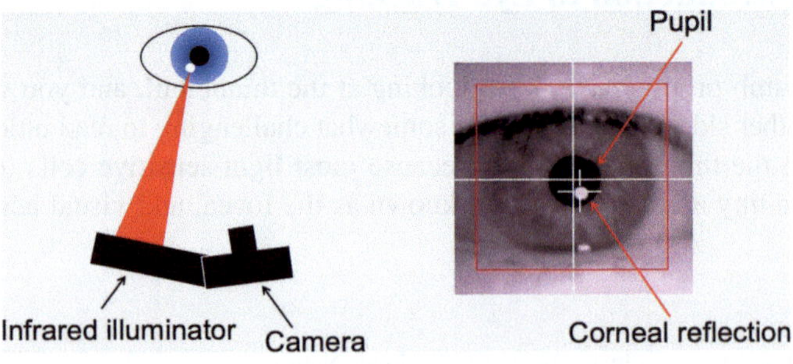

Fig. 4.1 The working principle of pupil-CR tracking

tracker is straightforward, and I would recommend users go through the relevant EyeLink user manual and learning resources (https://www.sr-support.com) for detailed guidelines and instructions.

The EyeLink trackers come with an application programming interface (API) that allows users to develop custom eye-tracking applications. The Python wrapper of this API is the Pylink library, and this chapter will focus on the fundamental functions it provides. The Pylink library is compatible with both Python 2 and 3. The example scripts included in this book have been tested with Python 3.6, but they should also work in more recent Python versions.

Installing Pylink

To use Pylink, you need first to install the latest version of the EyeLink Developer's Kit, which contains the EyeLink API and a few other useful tools. The EyeLink Developer's Kit supports all major platforms (Windows, macOS, and Ubuntu Linux). Windows and macOS installers for the EyeLink Developer's Kit are freely available from the SR Support Forum (https://www.sr-support.com/). For Ubuntu Linux, SR Research has set up a software repository for users and posted a detailed installation guide on the SR Support Forum. Once you have configured the software repository on your Ubuntu machine, you can run the following commands to install the EyeLink Developer's Kit.

```
sudo apt install eyelink-display-software
```

The "install_pylink.py" Script

Once the Developer's Kit is installed, the latest version of the Pylink library can be found in the following folder. This folder also contains an "install_pylink.py" script, which helps install Pylink for all versions of Python available on your computer.

Windows: C:\Program Files (x86)\SR Research\EyeLink\SampleExperiments\Python
macOS: /Applications/Eyelink/SampleExperiments/Python
Linux: /usr/share/EyeLink/SampleExperiments/Python

On macOS, run the following commands in the terminal. Please replace "python3.x" with the target version, e.g., "python3.6."

```
cd /Applications/Eyelink/SampleExperiments/python/
sudo python3.x install_pylink.py
```

The *install_pylink.py* script is also available once the EyeLink Developer's Kit is installed on Ubuntu Linux.

On Windows, open a command-line window by searching for "CMD" from the Start menu; then, execute the following commands (replace "-3.x" with the target version, e.g., "-3.6").

```
cd C:\Program Files (x86)\SR Research\EyeLink\SampleExperiments\Python\
py -3.x install_pylink.py
```

Manually Install Pylink

You can, of course, manually install Pylink by copying the Pylink library to the Python "site-packages" folder (Windows and macOS) or the "dist-package" folder (Ubuntu Linux). If unsure about where to find the "site-package" or "dist-package" folder, open a Python shell, and then type in the following commands to show the Python paths.

```
>>> import site
>>> site.getsitepackages()
```

On my MacBook, I have the following output in the Python shell.

```
Zhiguos-MacBook-Pro-2:~ zhiguo$ python3.8
Python 3.8.7 (v3.8.7:6503f05dd5, Dec 21 2020, 12:45:15)
[Clang 6.0 (clang-600.0.57)] on darwin
Type "help", "copyright", "credits" or "license" for more information.
>>> import site
>>> site.getsitepackages()
['/Library/Frameworks/Python.framework/Versions/3.8/lib/python3.8/
site-packages']
```

Once you have located the "site-package" folder, navigate to the folder that contains the "pylink" library, then copy the "pylink" library to the "site-packages" folder. On macOS or Linux, use the "cp" command with the "-R" option (copy recursively, i.e., copy the folder and its contents).

Manually installing Pylink is only needed if you use PsychoPy or Pygame as a Python module. The standalone version of PsychoPy has Pylink preinstalled; usually, no additional configuration is needed. Nevertheless, if you need to replace the

```
cd /Applications/Eyelink/SampleExperiments/python/
cp -R pylink /Library/Frameworks/Python.framework/Versions/3.8/lib/
python3.8/site-packages
```

version of Pylink included in the standalone version of PsychoPy, the correct folders to place the Pylink library are listed below.

Windows: C:\Program Files\PsychoPy3\Lib\site-packages
macOS: /Applications/PsychoPy.app/Contents/Resources/lib/python3.6

Installing Pylink Using pip

The Pylink library is also included in the EyeLink Developer's Kit as Python wheel files (in the "wheels" folder). Separate wheel files are provided for different versions of Python and architectures. You can use *pip* to install these locally available wheel files by executing a command in a Windows command prompt or a macOS/Ubuntu terminal (may require administrative privilege).

```
pip install *.whl
```

Please replace *.whl with a specific wheel file that matches the version of python and architecture you plan to use (see below for an example).

```
pip install sr_research_pylink-2.0.42917.0-cp38-cp38-macosx_10_9_
x86_64.whl
```

An Overview of An Eye-Tracking Experiment

The EyeLink eye-tracking system has a unique "two-computer" design that features a Host PC, dedicated to eye movement data recording, and a Display PC, responsible for stimulus presentation. The two computers communicate via an Ethernet link. This two-computer setup allows real-time data collection on the Host PC, which is not interrupted by the experimental task itself. The Ethernet link allows the Display PC to send over commands to control the eye tracker and retrieve real-time eye movement data if needed (e.g., for gaze-contingent tasks).

The Pylink library is a wrapper of core EyeLink API functions, which, among other things, allows users to open a connection to the Host PC, to transfer the camera image to the Display PC screen, and to calibrate the tracker on the Display

PC. We will give an overview of the frequently used Pylink functions in this chapter, but first, we will outline the operations typically involved in running an EyeLink experiment.

A typical experiment with an EyeLink tracker involves the following operations:

- Initialize a connection to the Host PC.
- Open an EDF (EyeLink Data Format) data file on the Host PC.
- Send commands to the Host PC to set tracking parameters.
- Open a full-screen window for camera setup and calibration.
- Calibrate the tracker and then run a set of experimental trials. Repeat this process if you need to test multiple blocks of trials.
- Close the EDF data file on the Host PC and transfer a copy to the Display PC, if needed.
- Terminate the connection to the Host PC.

We will begin with a very basic script containing just a few lines of code, while still covering all the operations listed above. Please bear in mind that this script is for illustration only; the later sections of this chapter will detail the additional commands required for an actual eye-tracking task. The code below is heavily commented on, and there is no need to explain the parameters of each of the functions called for now.

```
#!/usr/bin/env python3
#
# Filename: basic_example.py
# Author: Zhiguo Wang
# Date: 2/6/2021
#
# Description:
# A basic script showing how to connect/disconnect
# the tracker, open/close EDF data file, configure tracking parameter,
# calibrate the tracker, and start/stop recording

import pylink


# Step 1: Connect to the tracker
tk = pylink.EyeLink('100.1.1.1')

# Step 2: open an EDF data file on the EyeLink Host PC
tk.openDataFile('test.edf')

# Step 3: set some tracking parameters
tk.sendCommand("sample_rate 1000")
```

```
# Step 4: open a calibration window
pylink.openGraphics()

# Step 5: calibrate the tracker, then run five trials
tk.doTrackerSetup()

for i in range(5):
    # log a message in the EDF data file
    tk.sendMessage(f'Trial: {i}')

    # start recording
    tk.startRecording(1, 1, 1, 1)

    # record data for 2 seconds
    pylink.msecDelay(2000)

    # stop recording
    tk.stopRecording()

# Step 6: close the EDF data file and download it from the Host PC
tk.closeDataFile()
tk.receiveDataFile('test.edf', 'test.edf')

# Step 7: close the link to the tracker, then close the window
tk.close()
pylink.closeGraphics()
```

Connect/Disconnect the Tracker

At the beginning of an eye-tracking experiment, we need to establish an active connection to the Host PC. Without this connection, the experimental script cannot send over commands to control the tracker, nor can it receive gaze data from the eye tracker. The command for initializing a connection is *pylink.EyeLink()*. This command takes just one parameter, the IP address of the Host PC. If you omit the IP address, this function will use the default address of the EyeLink Host PC, which is 100.1.1.1.

The command *pylink.EyeLink()* returns an EyeLink object (i.e., *tk* in the example code below), which has a set of methods that we can use to interact with the eye tracker via the software running on the Host PC. Once there is an active connection, the EyeLink Host PC status will switch to "Link Open" (shown in the top-right corner of its screen).

```
tk = pylink.EyeLink('100.1.1.1')
```

To communicate with the Host PC, the Display PC has to be on the same network as the Host PC. Following the EyeLink user manual, the Display PC IP address is typically set to 100.1.1.2, and the subnet mask is 255.255.255.0. The most common cause of "Connection Failed" errors is not setting the IP address of the Display PC correctly.

The eye tracker in your lab may not always be available, or you may find it more convenient to debug your experimental script on a computer that is not physically connected to the EyeLink Host PC. If your script does not rely on real-time eye movement data, you can open a simulated connection to the tracker with the following command.

```
tk = pylink.EyeLink(None)
```

When a task finishes, we terminate the connection with the *close()* command.

```
tk.close()
```

Open/Close EDF Data File

The EyeLink Host PC is a machine dedicated to eye tracking and data logging. At the beginning of a new testing session, we open a data file on the Host PC to store the eye movement data. The data file can be retrieved from the Host PC at the end of a testing session after closing the data file. For backward compatibility, the EDF file name should not exceed eight characters (not including the ".edf" extension). To open an EDF data file on the Host PC, use the following command.

```
tk.openDataFile('test.edf')
```

To close an EDF data file currently open on the Host PC, call the following command.

```
tk.closeDataFile()
```

Configure the Tracker

Various parameters (e.g., sampling rate or eye to track) can be changed by clicking the relevant GUI buttons on the EyeLink Host PC software. However, a less error-prone approach is to configure tracker parameters by sending commands to the Host PC from your experimental script, with the *sendCommand*() function. For instance, the command below sets the sample rate to 1000 Hz.

```
tk.sendCommand("sample_rate 1000")
```

Open a Window for Calibration

As noted, we need to calibrate the tracker to estimate which spatial location the participant is looking at. For a screen-based task, calibration usually involves the presentation of a gaze target at different screen locations. As we have seen in the earlier chapters, the first step when drawing things to the screen is to open a window. The most straightforward approach is to call the Pylink function *pylink.openGraphics()* to open a calibration window. But a PsychoPy or Pygame window can also be used for calibration (see Chap. 7 for information on custom calibration graphics).

```
pylink.openGraphics()
```

Call the following command to close the window when tracker calibration completes or a testing session ends.

```
pylink.closeGraphics()
```

Calibrate the Tracker

With the EyeLink tracker, calibration is typically a two-step process: calibrating the tracker and evaluating the resulting calibration model. Both steps involve presenting a series of visual targets at different known screen positions. The observer is required to shift gaze to follow the calibration target. The calibration process can use 3, 5, 9, or 13 screen positions. A 9-point calibration will give you the best results if you use a chin rest to stabilize the head of the observer. A 5-point or 13-point calibration will provide you with better results when the head is free to move (i.e., tracking in

Remote Mode). Following calibration, there is a validation process in which a visual target appears at known screen positions, and the observer shifts gaze to follow the target. By comparing the gaze position reported by the tracker and the physical position of the target, the tracker can estimate the gaze errors at different screen positions. This two-step procedure (calibration and validation) requires a single Pylink command.

```
tk.doTrackerSetup()
```

In addition to presenting the calibration targets, the *doTrackerSetup*() command can also transfer the camera image to the Display PC (see Fig. 4.2), to allow for camera focusing and threshold adjustments. Once the tracker has entered the calibration mode, the experimenter can press Enter to show or hide the camera image (see Fig. 4.2, left panel), press C to calibrate, press V to validate, and press O (short for output) to exit the calibration routine.

Start/Stop Recording

It is generally a good idea to start data recording at the beginning of each trial and stop recording at the end of each trial. By doing so, the tracker skips the inter-trial intervals, reducing the size of the EDF data file and simplifying things at the analysis stage. For situations where a single continuous recording is preferred (e.g., in an fMRI or EEG study), data recording can be started at the beginning of a session (or run) and stopped at the end of a session.

To start data recording, call *startRecording().* This command takes four parameters specifying what types of data (event or sample; see Chaps. 6 and 8 for a brief discussion on EyeLink data types) are recorded in the EDF data file and what types of data are available over the link during testing. With four 1's, the tracker will

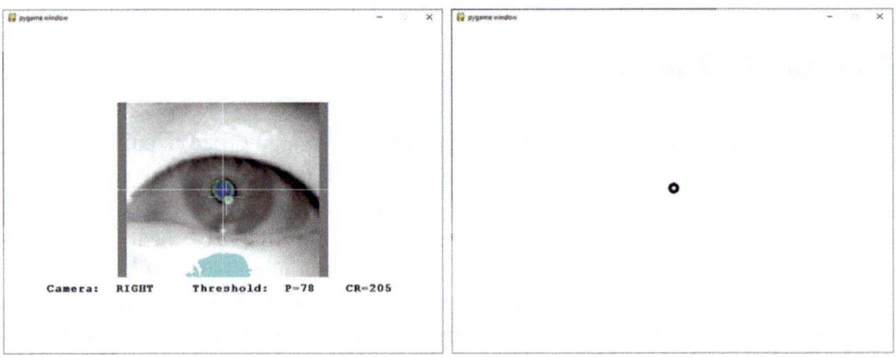

Fig. 4.2 A calibration window showing the camera image (left) and the calibration target (right)

record both events and samples in the data file and also make these two types of data available over the link.

To stop recording, call the *stopRecording*() command.

```
tk.stopRecording()
```

Log Messages

It is recommended to log messages to mark the various events that occurred during testing and to store information that would facilitate data analysis. In the example script, the *sendMessage*() command is called to log the current trial number. All messages are timestamped and stored in the EDF data file.

```
    tk.sendMessage(f'Trial: {i}')
```

Retrieve the EDF Data File

All EDF data files are stored on the Host PC unless they are deleted or overwritten. When a testing session finishes, we close the EDF file with *closeDataFile*() and download the EDF file to the Display PC with *receiveDataFile*(). The *receiveDataFile*() command takes two parameters—the EDF file you would like to download and the place to store the EDF file on the Display PC. The command below does not alter the EDF file name; you can, of course, rename the downloaded EDF file and add it to the desired folder (e.g., "~/results/subj_01.edf"), if needed.

```
tk.closeDataFile()
tk.receiveDataFile('test.edf', 'test.edf')
```

A Real Example: Free Viewing

The example script in the previous section illustrates the critical steps in eye-tracker integration but does not present any stimuli to the participants. This section will give a complete script to illustrate the frequently used commands and the scripting protocol recommended for the EyeLink trackers. The task implemented in this example script is straightforward. We present an image on the screen, and the participant presses a key when he/she finishes viewing the image. The complete script is listed below; I will explain the critical bits of the script in detail.

The Pygame library is used to show the visual stimuli and register participant responses in this script. The reason is that the built-in Pylink function *pylink.openGraphics*() is based on the SDL library, the same library behind Pygame. So, Pylink can use an existing Pygame window to calibrate the tracker when we call *pylink.openGraphics*(). In later chapters, we will discuss how to use a PsychoPy window for calibration and how to customize the behavior of the calibration routine.

```
#!/usr/bin/env python3
#
# Filename: free_viewing.py
# Author: Zhiguo Wang
# Date: 2/7/2020
#
# Description:
# A free-viewing task implemented in Pygame.
# Press any key to terminate a trial

import os
import random
import pylink
import pygame
from pygame.locals import *

# Screen resolution
SCN_W, SCN_H = (1280, 800)

# Step 1: Connect to the tracker
tk = pylink.EyeLink('100.1.1.1')

# Step 2: open an EDF data file on the EyeLink Host PC
tk.openDataFile('freeview.edf')
# Optional file header
tk.sendCommand("add_file_preamble_text 'Free Viewing Task'")

# Step 3: Set tracking parameters, e.g., sampling rate
#
# Put the tracker in offline mode before we change its parameters
tk.setOfflineMode()

# Set the sampling rate to 1000 Hz
tk.sendCommand("sample_rate 1000")

# Send screen resolution to the tracker
tk.sendCommand(f"screen_pixel_coords = 0 0 {SCN_W-1} {SCN_H-1}")
```

```
# Record a DISPLAY_SCREEN message to let Data Viewer know the
# correct screen resolution to use when visualizing the data
tk.sendMessage(f'DISPLAY_COORDS 0 0 {SCN_W - 1} {SCN_H - 1}')

# Set the calibration type to 9-point (HV9)
tk.sendCommand("calibration_type = HV9")

# Step 4: open a Pygame window; then, call pylink.openGraphics()
# to request Pylink to use this window for calibration
pygame.display.set_mode((SCN_W, SCN_H), DOUBLEBUF | FULLSCREEN)
pygame.mouse.set_visible(False)  # hide the mouse cursor
pylink.openGraphics()

# Step 5: calibrate the tracker, then run through the trials
tk.doTrackerSetup()

# Parameters of all trials stored in a list
t_pars = [
    ['quebec.jpg', 'no_people'],
    ['woods.jpg', 'with_peole']
    ]

# Define a function to group the lines of code that will be executed
# in each trial
def run_trial(params):
    ''' Run a trial

  params: image, condition in a list,
    e.g., ['quebec.jpg', 'no_people'] '''

    # Unpacking the picture and correct keypress

  pic, cond = params

    # Load the picture, scale the image to fill up the screen

  pic_path = os.path.join('images', pic)
    img = pygame.image.load(pic_path)
    img = pygame.transform.scale(img, (SCN_W, SCN_H))

    # Record_status_message: show some info on the Host PC

  tk.sendCommand(f"record_status_message 'Picture: {pic}'")

    # Drift-check; re-calibrate if ESCAPE is pressed
    # parameters: x, y, draw_target, allow_setup

  tk.doDriftCorrect(int(SCN_W/2), int(SCN_H/2), 1, 1)
```

```
# Start recording
# parameters: file_event, file_sample, link_event, link_sample
tk.startRecording(1, 1, 1, 1)
# Wait for 100 ms to cache some samples
pylink.msecDelay(100)

# Present the image
surf = pygame.display.get_surface()
surf.blit(img, (0, 0))
pygame.display.flip()

# Log a message to mark image onset
tk.sendMessage('image_onset')

# Log a '!V IMGLOAD' message to the EDF data file, so Data Viewer
# knows where to find the image when visualizing the gaze data
img_path = f'images/{pic}'
tk.sendMessage(f'!V IMGLOAD FILL {img_path}')

# Wait for a key response
pygame.event.clear()  # clear all cached events
got_key = False
while not got_key:
    for ev in pygame.event.get():
        if ev.type == KEYDOWN:
            tk.sendMessage(f'Keypress {ev.key}')
            got_key = True

# clear the screen
surf.fill((128, 128, 128))
pygame.display.flip()

# clear the host backdrop as well
tk.sendCommand('clear_screen 0')

# Log a message to mark image offset
tk.sendMessage('image_offset')

# stop recording
tk.stopRecording()
```

```
# Run through the trials in a random order
random.shuffle(t_pars)
for trial in t_pars:
   run_trial(trial)

# Step 6: close the EDF data file and download it
tk.closeDataFile()
tk.receiveDataFile('freeview.edf', 'freeview.edf')

# Step 7: close the link to the tracker and quit Pygame
tk.close()
pygame.quit()
```

As noted, this script uses Pygame functions to present the pictures and to register keyboard responses. So, the script starts with importing Pygame and the Pygame local constants (see Chap. 3). We then specify the screen dimensions by unpacking a two-item tuple (1280, 800) onto two variables, *SCN_W* and *SCN_H*. We store the parameters of all trials in a list (*t_pars*), just like we did with the example scripts in Chaps. 2 and 3. For the commands relevant to eye tracking, I will highlight a few important ones here.

Preamble Text in EDF File

It is a good practice to put some header information in the EDF data file. Otherwise, it is difficult to tell which EDF data file belongs to which research project. Please note that the header text, i.e., "Free Viewing Task," needs to be in a pair of single quotes and the entire command needs to be in double quotes. Make sure that this is the first EyeLink command/message sent in the experiment session.

```
# optional file header to identify the experimental task
tk.sendCommand("add_file_preamble_text 'Free Viewing Task'")
```

Offline Mode

The *setOfflineMode()* command is frequently called in EyeLink scripts. This command will put the tracker into the offline (idle) mode, so it is ready to accept configuration commands.

EyeLink Host Commands

A variety of commands can be sent by Pylink to control the tracker. This example script illustrates a few frequently used commands, e.g., "sample_rate." The "screen_pixel_coords" command is particularly important, as the tracker needs this information to estimate eye movement velocity in degrees of visual angle. The velocity data is necessary for accurate online detection of eye events (saccades and fixations). The "calibration_type" command controls the size of the calibration grid, e.g., 3-point (HV3), 5-point (HV5), and 9-point (HV9).

```
# Set the sampling rate to 1000 Hz
tk.sendCommand("sample_rate 1000")

# Send screen resolution to the tracker
tk.sendCommand(f"screen_pixel_coords = 0 0 {SCN_W-1} {SCN_H-1}")

# Set the calibration type to 9-point (HV9)
tk.sendCommand("calibration_type = HV9")
```

Calibration Window

As noted, the *pylink.openGraphics()* command will open a window so that the application can draw the eye-tracker camera image, display calibration target, and handle keyboard events on the Display PC. There is no need to discuss all the routines encapsulated in this function right now. If there is already an active Pygame window, Pylink will use it for calibration when we call *pylink.openGraphics()*.

Record Status Message

At the beginning of each trial, it is recommended to show some trial information in the bottom-right corner on the Host PC screen (see Fig. 4.3). This is achieved by sending over a "record_status_message" command.

```
# Record_status_message: show some info on the Host PC
tk.sendCommand(f"record_status_message 'Picture: {pic}'")
```

In this example script, the record status message shows the current trial number only. It can be more informative, so the experimenter can easily tell which trial is being tested, how many trials have been completed, which experimental condition the current trial belongs to, etc. (Fig. 4.3).

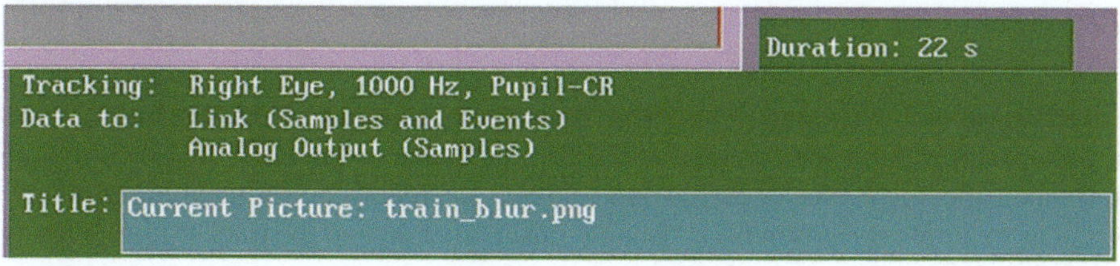

Fig. 4.3 Recording status message is shown in the gray box in the bottom-right corner on the EyeLink Host PC

Drift Check/Drift Correction

One of the most critical commands illustrated in this example script is *doDriftCorrect()*. Following this command, a target appears on the screen, and the experimenter (or the participant) is required to look at the target. The experimenter (or the participant) presses a key to register the current gaze position. The tracker will compare the gaze position estimated by the tracker and the physical position of the target and report the current tracking accuracy.

This feature is known as drift correction or drift check, and one can think of it as a 1-point validation of tracking accuracy. The concept of drift correction is inherited from earlier head-mounted versions of the EyeLink eye tracker. In these models, the headband may slip and cause drift in the gaze data (especially when tracking in pupil-only mode). This issue was tackled by a linear correction of the gaze data based on the gaze error reported by the drift-correction procedure. For recent EyeLink eye trackers, by default, any gaze error at the drift-check target is no longer used to correct the calibration model, as the pupil-CR tracking algorithm is resilient to small head or camera displacement. Instead, for all recent EyeLink models, the drift-correction routine acts as a "drift check"; it checks the tracking accuracy and allows users to recalibrate if necessary.

```
# Drift-check; re-calibrate if ESCAPE is pressed
# parameters: x, y, draw_target, allow_setup
tk.doDriftCorrect(int(SCN_W/2), int(SCN_H/2), 1, 1)
```

The *doDriftCorrect*() command takes four parameters. The first two are x, y pixel coordinates for the drift-check target. Note that x, y must be integers, e.g., 512, 384. The third parameter specifies whether Pylink should draw the target. If set to 0, we would need to first draw a custom target at the (x, y) pixel coordinates and then call the *doDriftCorrect*() command. The fourth parameter controls whether Pylink should evoke the calibration routine if the ESCAPE key is pressed.

Log Messages

Messages should be sent to the tracker whenever a critical event occurs; for instance, a picture appears on the screen. With these messages in the EDF data file, we can tell when specific events occurred and segment the recording for meaningful analysis. The example script sends an "Image_onset" message to the tracker immediately after the picture appears on the screen.

```
# send a message to mark image onset
tk.sendMessage('Image_onset')
```

In addition to user-defined messages, we can also log messages useful for offline data analysis and visualization in SR Research Data Viewer. We will discuss these messages in considerable detail in Chap. 5. In this basic example script, the special "!V" prefix to the IMGLOAD message indicates that the message is intended for Data Viewer.

```
# Record a DISPLAY_SCREEN message to let Data Viewer know the
# correct screen resolution to use when visualizing the data
tk.sendMessage(f'DISPLAY_SCREEN 0 0 {SCN_W - 1} {SCN_H - 1}')

# Log a '!V IMGLOAD' message to the EDF data file, so Data Viewer
# knows where to find the image when visualizing the gaze data
img_path = os.path.join('images', pic)
tk.sendMessage(f'!V IMGLOAD FILL {img_path}')
```

Error and Exception Handling

One important topic that has not been covered so far is error and exception handling, i.e., what should we do if a command fails. To keep the code concise and easy to follow, error handling is not fully implemented in the example scripts presented in this book. This does not mean that error handling is not important. Quite the contrary. To show the importance of error handling, I will give two examples relevant to the short free-viewing script presented in the previous section.

If you execute the *pylink.EyeLink('100.1.1.1')* command when the tracker is not connected, the Python interpreter will raise a Runtime Error, which complains that it "Could not connect to tracker at 100.1.1.1." This error will crash your script and print out lots of other error messages.

```
>>> pylink.EyeLink('100.1.1.1')
Connection timed outTraceback (most recent call last):
  File "<stdin>", line 1, in <module>
   File "/Library/Frameworks/Python.framework/Versions/3.6/lib/python3.6/
site-packages/pylink/eyelink.py", line 207, in __init__
    self.open(trackeraddress,0)
RuntimeError: Could not connect to tracker at 100.1.1.1
```

A proper way to handle this error is to put the *pylink.EyeLink('100.1.1.1')* command in a *try* statement.

```
# Connect to the tracker
try:
    tk = pylink.EyeLink('100.1.1.1')
except RuntimeError as err:
    fix = '\nDoulbe-check the IP address of the stimulus presentation PC!'
    print(err, fix)
```

With the try statement, instead of a hard crash, the Python interpreter will give a customized warning message.

```
>>>
Could not connect to tracker at 100.1.1.1
Double-check the IP address of the stimulus presentation PC!
```

I will not delve into the details, but interested users should refer to the official Python documentation (https://docs.python.org/3/tutorial/errors.html). A similar trick can be used to handle exceptions related to the *doDriftCorrect*() command in the free-viewing script; for instance, if this command fails, we call *doTrackerSetup*() to recalibrate the tracker.

In addition to the built-in Python routines for exception handling, custom code can also be used to handle potential errors. For instance, the EDF data filename can contain only letters, numbers, and underscore ("_"), and there is a length limit of eight characters. If you need to specify a unique EDF data filename for each testing session, it is wise to check whether the filename follows these rules.

```
from string import ascii_letters, digits

# Loop until a valid EDF data filename is specified
print('Specify an EDF data filename. \n' + \
'The filename should contain only letters, numbers\n' + \
'and underscore; it should not exceed 8 characters')
while True:
    edf_fname = input('\n-->')

# strip trailing spaces, if there were any
edf_fname = edf_fname.rstrip()

# remove the '.edf' extension if it is in 'edf_fname'
edf_fname = edf_fname.split('.')[0]

# check if the filename is valid
allowed_char = ascii_letters + digits + '_'
if not all([c in allowed_char for c in edf_fname]):
     print('ERROR: Invalid character included in the filename')
elif len(edf_fname) > 8:
     print('ERROR: EDF filename should not exceed 8 characters')
else:
     break

# print the EDF filename
print('The filename you specified is:', edf_fname)
```

With this code snippet, a Python shell will come up, prompting you to specify an EDF data filename. The Python interpreter will only break the loop when you have entered a valid filename.

```
>>>
Specify an EDF data filename.
The filename should contain only letters, numbers
and underscore; it should not exceed 8 characters

-->asdfg@
ERROR: Invalid character included in the filename

-->abcd12345
ERROR: EDF filename should not exceed 8 characters

-->abcd1234.edf
The filename you specified is:  abcd1234
```

As noted, for brevity, error handling is not fully considered in the example scripts presented in this book. I would strongly recommend readers to take a look at the example scripts included in the EyeLink Developer's Kit to understand good practices in this regard. Errors and exceptions that may occur when using the Pylink library are properly handled in these scripts.

In this chapter, we briefly discussed eye tracking and the basic features and functions of the Pylink library. When programming a task, it is crucial to bear in mind data analysis and visualization needs. A small amount of time invested in the experimental script can save a great deal of time at the analysis stage. Chapter 5 will discuss the programming protocols that allow seamless integration with Data Viewer, a powerful data analysis and visualization software provided by SR Research.

Chapter 5
Preparing Scripts that Support Analysis and Visualization in Data Viewer

Contents

Data Viewer is a software package developed by SR Research to visualize and analyze EyeLink® data (see Fig. 5.1). It is, of course, possible to write a Python script to retrieve the eye movement data needed for a particular analysis (see Chap. 8). Still, Data Viewer is a very convenient tool for processing eye gaze data recorded with EyeLink eye trackers. Frequently used dependent measures, such as dwell time and fixation count for interest areas, are just a few mouse clicks away.

When analyzing gaze data, it is usually preferable to segment the recordings into "trials," i.e., short periods during which we manipulate a particular stimulus or experimental condition. An experiment usually contains tens, if not hundreds, of trials, allowing the researcher to estimate the typical (average) pattern of results in a given experimental condition. To allow Data Viewer to segment the gaze data into trials and facilitate further analysis, we need to record some information in the EyeLink data file so that Data Viewer knows (a) the start and end time of each trial, (b) which condition or conditions a trial belongs to, (c) the onset/offset of critical trial events such as stimulus onset, (d) the location and onset time of the interest areas associated with a trial, and (e) the participant's responses in a trial (response time, accuracy, etc.). It is also helpful to have the images or videos we presented

Z. Wang, *Eye-Tracking with Python and Pylink*,
https://doi.org/10.1007/978-3-030-82635-2_5

Fig. 5.1 The Data Viewer software developed by SR Research is a convenient tool for analyzing and visualizing eye movement data

during the task available to Data Viewer. Gaze data can then be plotted over the background images or videos to facilitate data analysis and interpretation.

In an experimental script, special "Data Viewer Integration" messages can be recorded in the eye movement data file to facilitate data analysis in Data Viewer. The following sections will discuss the various Data Viewer integration messages and illustrate their usage with short scripts. These scripts are for illustration purposes. If you need a template for your experimental task, use the example script presented in the final section of this chapter. You may also refer to the EyeLink Data Viewer User Manual for detailed description of integration messages.

Trial Segmentation

Data Viewer needs to know the start and end time of each trial to segment recordings into trials, much like epoching in EEG data analysis. The beginning and end of a trial are marked by messages that start with reserved keywords TRIALID and TRIAL_RESULT. When Data Viewer sees the keywords "TRIALID" and "TRIAL_RESULT," it knows they mark the start and end of a trial. It is not mandatory, but you may include additional information in the TRIALID and TRIAl_RESULT messages.

```
TRIALID <values that help to identify a trial>
TRIAL_RESULT <values representing possible trial result>
```

Messages are written to the EDF file using the *sendMessage*() function. Assume "tk" is the tracker connection we have initialized; we can include the following line of code at the beginning of each trial. You will always need to send some data following the TRIALID token to help identify the current trial.

```
tk.sendMessage('TRIALID 1')
```

At the end of a trial, we include the following line of code to send a "TRIAL_RESULT" message. Here, we add a flag "0" to the TRIAL_RESULT message to indicate the trial completed successfully.

```
tk.sendMessage('TRIAL_RESULT 0')
```

The following script illustrates the use of TRIALID and TRIAL_RESULT messages in a typical experiment.

```
#!/usr/bin/env python3
#
# Filename: trial_segmentation.py
# Author: Zhiguo Wang
# Date: 2/6/2021
#
# Description:
# This script illustrates the TRIALID and TRIAL_RESULT messages
# that Data Viewer uses to segment a recording into trials

import pylink


# Connect to the tracker
tk = pylink.EyeLink()

# Open an EDF on the Host
# filename must not exceed 8 characters
tk.openDataFile('seg.edf')

# Run through five trials
for trial in range(1, 6):
    # Print out a message to show the current trial
    print(f'Trial #: {trial}')
```

```
        # Log a TRIALID message to mark trial start
        tk.sendMessage(f'TRIALID {trial}')

        # Start recording
        tk.startRecording(1, 1, 1, 1)

        # Pretending that we are doing something for 2-sec
        pylink.pumpDelay(2000)

        # Stop recording
        tk.stopRecording()

        # Log a TRIAL_RESULT message to mark trial ends
        tk.sendMessage('TRIAL_RESULT 0')

    # Wait for 100 to catch session end events
    pylink.msecDelay(100)

    # Close the EDF file and download it from the Host PC
    tk.closeDataFile()
    tk.receiveDataFile('seg.edf', 'trial_segmentation_demo.edf')

    # Close the link
    tk.close()
```

Here we used an f-string to embed the current trial number (1, 2, 3, etc.) in the TRIAL_ID message. There will be five trials when we load the resulting data file into Data Viewer. As is clear from Fig. 5.2, these five trials are correctly segmented, with the TRIALID and TRIAL_RESULT messages marking the beginning and end of each trial.

As mentioned previously, in most EyeLink experiments, the eye tracker is instructed to start and stop recording for each trial. We send the TRIALID message before the recording starts and the TRIAL_RESULT message after the recording ends. However, this is not mandatory; in fMRI and EEG studies, you may send the TRIALID and TRIAL_RESULT messages to the tracker while data recording is ongoing. In this way, you get a single continuous recording, which contains multiple pairs of TRIALID and TRIAL_RESULT messages that allow the Data Viewer software to segment the recording into trials.

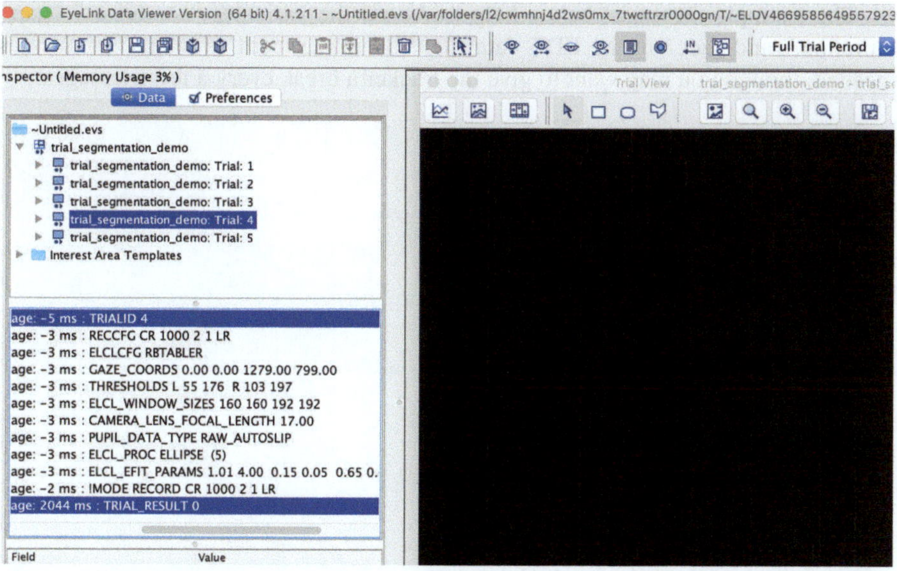

Fig. 5.2 Data Viewer, by default, uses the keyword "TRIALID" and "TRIAL_RESULT" to segment the eye movement recordings into trials

Trial Variables

An experiment usually involves the manipulation of variables. For instance, in a Stroop task, we have trials on which the color of a word is congruent or incongruent with the meaning of the same word; for instance, showing the word "RED" in blue would give us an incongruent trial. It is generally good to record all variables manipulated in a task in the EDF data file. We can then group the eye movement data based on these variables for visualization or statistical analysis. In addition to the variables manipulated by the experimenter, a trial may give rise to additional variables such as response time and accuracy. These variables can also be helpful for eye movement data analysis. For instance, trials can be grouped by correct and incorrect responses, or trials with extremely fast or slow responses can be excluded.

Trial variable messages should generally be sent at the end of each trial, allowing variables that may get updated during the recording, such as response time, accuracy, etc., to be included. These messages should be sent before the TRIAL_RESULT message, which marks the end of a trial. The format of a trial variable message is listed below. The prefix "!V" in the message informs Data Viewer that the message is for a particular purpose. Following the prefix are the keyword "TRIAL_VAR" and two additional parameters—<trial_var_label> and <trial_var_value>, i.e., the name of the variable and the value we assign to the variable in the current trial.

```
!V TRIAL_VAR <trial_var_label > <trial_var_value >
```

The example messages below help to record the variables "condition," "gap_duration," and "acc" in the EDF data file. You may record as many variables as needed, though you may want to give the tracker a break every a few messages.

```
# Run through five trials
for trial in range(1, 6):
    # Print out a message to show the current trial
    print(f'Trial #: {trial}')

    # Log a TRIALID message to mark trial start
    tk.sendMessage(f'TRIALID {trial}')

    # Start recording
    tk.startRecording(1, 1, 1, 1)

    # Pretending that we are doing something for 2-sec
    pylink.pumpDelay(2000)

    # Stop recording
    tk.stopRecording()

    # Send TRIAL_VAR messages to store variables in the EDF
    tk.sendMessage('!V TRIAL_VAR condition step')
    tk.sendMessage('!V TRIAL_VAR gap_duration 200')
    tk.sendMessage('!V TRIAL_VAR direction Right')

    # Log a TRIAL_RESULT message to mark trial ends
    tk.sendMessage('TRIAL_RESULT 0')
```

After loading the EDF data file into Data Viewer, you can examine the recorded trial variables in the Trial Variable Value Editor. The timestamps of the trial variable messages do not matter. The variables will be correctly parsed by Data Viewer as long as they are logged between the TRIALID and TRIAL_RESULT messages. The values of the variables did not vary from trial to trial in this code snippet. This is seldom the case in an actual experiment. In the final section of this chapter, we will present a few complete experimental scripts that you can use as templates. These examples illustrate how to change the variable values appropriately for each trial (Fig. 5.3).

Interest Areas

Interest areas (IAs), also known as areas of interest (AOIs) or regions of interest (ROIs), are indispensable in many gaze data analyses. In Data Viewer, an IA can be rectangular, elliptic, or a "freehand" shape. Each IA has a unique ID (e.g., 1, 2, 3)

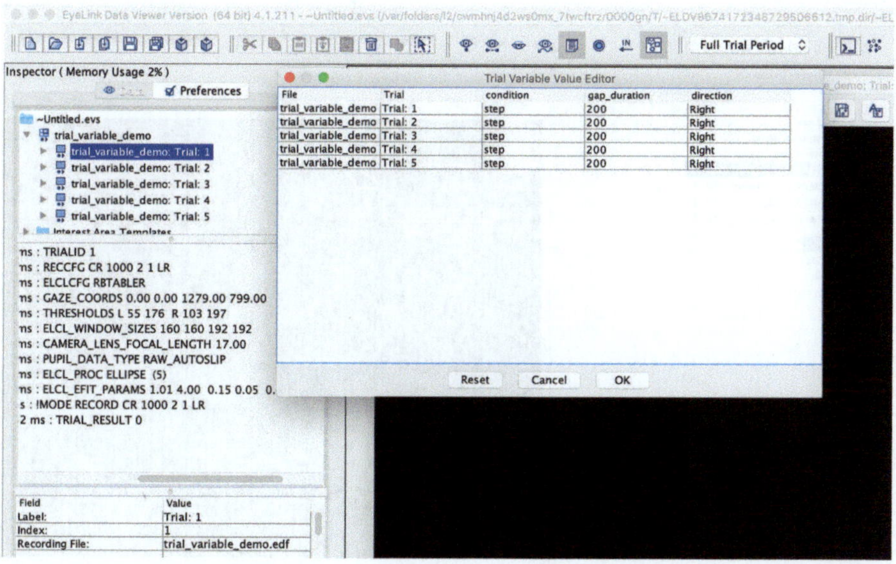

Fig. 5.3 Data Viewer automatically parses the trial variable messages when importing EDF data files. The variables and values can be viewed or edited in the Trial Variable Value Editor

and a meaningful label (optional). While interest areas can be defined manually in Data Viewer after data collection, they can also be defined in experimental scripts. Creating IAs in the experimental script itself has several advantages. It is particularly appropriate for tasks such as Visual Search or Visual World paradigm in which the location of the target and distractors may change across trials. Special "Interest Area" messages can be written into the EDF files, and these messages will allow Data Viewer to display the IAs during analysis. The interest area messages logged during testing can take one of the following formats. The message always starts with the prefix "!V," followed by the "IAREA" command and a flag specifying the shape of the interest area, i.e., "RECTANGLE," "ELLIPSE," and "FREEHAND." The interest area <id> option is mandatory, and it should be unique for each interest area, but the interest area [label string] is optional.

```
!V IAREA RECTANGLE <id> <left> <top> <right> <bottom> [label string]
!V IAREA ELLIPSE <id> <left> <top> <right> <bottom> [label string]
!V IAREA FREEHAND <id> <x1, y1> <x2, y2 > ... <xn, yn> [label string]
```

For an illustration, we insert the following interest area messages into the minimalist-style script we have been using so far, right after the *tk.startRecording(1, 1, 1, 1)* command. Bear in mind that Data Viewer will use the timestamps of the interest area messages to determine when to show them during trial playback. In Data Viewer, the details of each interest area can be examined by selecting the "Custom Interest Area Set" field under each trial (see Fig. 5.4).

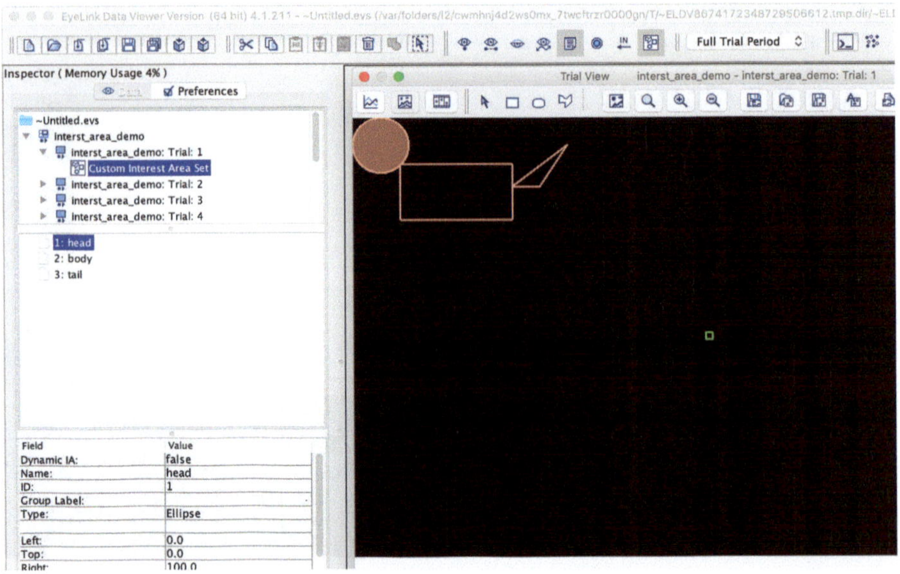

Fig. 5.4 Data Viewer reconstructs the interest areas from the messages in the EDF data files

Suppose the experimental task involves many interest areas (e.g., a reading task). In that case, all of the interest area definitions can be placed in a single plain text file, known as Interest Area Set (.ias) file. In the .ias file, multiple interest areas are

```
# Run through five trials
for trial in range(1, 6):
    # Print out a message to show the current trial
    print(f'Trial #: {trial}')

    # Log a TRIALID message to mark the trial start
    tk.sendMessage(f'TRIALID {trial}')

    # Start recording
    tk.startRecording(1, 1, 1, 1)

    # Send IAREA messages to store interest area definitions
    tk.sendMessage('!V IAREA ELLIPSE 1 0 0 100 100 head')
    tk.sendMessage('!V IAREA RECTANGLE 2 85 85 285 185 body')
    tk.sendMessage('!V IAREA FREEHAND 3 285,125 385,50 335,125 tail')

    # Pretending that we are doing something for 2-sec
```

```
pylink.pumpDelay(2000)

# Stop recording
tk.stopRecording()

# Log a TRIAL_RESULT message to mark trial ends
tk.sendMessage('TRIAL_RESULT 0')
```

defined, with each in a single line. Note "!V" is no longer needed for the interest area definitions in a .ias file.

```
RECTANGLE   1   51    29   146   82   Buck
RECTANGLE   2   146   29   195   82   did
RECTANGLE   3   195   29   244   82   not
RECTANGLE   4   244   29   311   82   read
RECTANGLE   5   311   29   360   82   the
RECTANGLE   6   360   29   536   82   newspapers
```

To point Data Viewer to read in an interest area set file, include a message in the following format in the experimental script.

```
!V IAREA FILE <ias_file>
```

The interest area set files may also contain the start and end timestamps of each interest area, which can be helpful in creating "dynamic" interest areas. Please see the Data Viewer user manual for more information on the format of interest area.

Background Graphics

For data inspection and visualization, it is helpful to have the background graphics (e.g., images) and the gaze data (samples, fixations, saccades, etc.). As the EDF file only contains eye-tracking data and messages but not the images, the desired solution is to embed messages in the EDF data files that point to where the images are stored. In that way, when importing the data files, Data Viewer will find the background graphics for each trial. The background graphics can be images, simple drawings, or even videos.

As noted in Chap. 4, it is crucial to let the tracker know the screen dimension (in pixels), so the tracker can use this information to estimate the velocity of the eye movements. To correctly overlay gaze on top of the background graphics, we must also log a DISPLAY_COORDS message in the EDF data file. This message informs Data Viewer of the size of the screen, and its format is listed below.

```
DISPLAY_COORDS <left> <top> <right> <bottom>
```

This message is usually sent to the tracker before data recording begins. In the example code below, the DISPLAY_COORDS message is a reserved keyword and does not have the "!V" prefix, as with the TRIALID and TRIAL_RESULT messages.

```
tk.sendMessage('DISPLAY_COORDS 0 0 1023 767')
```

Images

For messages specifying the background images, the format can be one of the following. The command "!V IMGLOAD" is followed by a keyword, which specifies the reference point for the image. The "FILL" command will stretch the image to fill the whole screen; the "TOP_LEFT" and "CENTER" commands tell Data Viewer to present the top-left corner or the center of the image at the specified screen pixel coordinates, i.e., <x_position> and <y_position>. The [width] and [height] parameters are optional.

```
!V IMGLOAD FILL < image >
!V IMGLOAD TOP_LEFT <image > <x_position> <y_position> [width] [height]
!V IMGLOAD CENTER <image > <x_position> <y_position> [width] [height]
```

The <image> file name should contain the full path to the image file, relative to the location of the EDF data file itself. Suppose our script folder has the following structure, and the EDF data files are saved in the "data_files" folder, whereas the images used by the script are saved in the "images" folder.

The message we send to the tracker immediately after "*image_1.png*" appears on the screen should be in the following format. In this message, "." (one dot) represents the EDF data folder (i.e., "data_files"), and ".." (two dots) means to move up one level in the folder hierarchy.

```
tk.sendMessage('!V IMGLOAD FILL ./../images/image_1.png')
```

```
Script_folder
      |__experiment_script.py
      |__images
      |          |__ image_1.png
      |          |__ image_2.png
      |__data_files
                 |__test.edf
```

To illustrate the usage of the IMGLOAD command, we add a message immediately after recording start in the script we have been using so far. The format of this message is much simpler than the above one as the image is stored in the same directory as the EDF data file. Here we also send the DISPLAY_COORDS command to record the correct screen resolution to use for visualization.

```
#!/usr/bin/env python
#
# Filename: image_load.py
# Author: Zhiguo Wang
# Date: 2/6/2021
#
# Description:
# This script illustrates the IAREA messages
# that Data Viewer uses to reconstruct interest areas

import pylink


# Connect to the tracker
tk = pylink.EyeLink()

# Open an EDF on the Host; filename must not exceed 8 characters
tk.openDataFile('imgload.edf')

# Assume the screen resolution is 1280 x 800 pixels
SCN_W, SCN_H = (1280, 800)

# Pass the screen coordinates to the tracker
coords = f"screen_pixel_coords = 0 0 {SCN_W - 1} {SCN_H - 1}"
tk.sendCommand(coords)
```

```
# Record a DISPLAY_COORDS message to let Data Viewer know the
# correct screen resolution to use when visualizing the data
tk.sendMessage(f'DISPLAY_COORDS 0 0 {SCN_W - 1} {SCN_H - 1}')

# Run through five trials
for trial in range(1, 6):
    # Print out a message to show the current trial
    print(f'Trial #: {trial}')

    # Log a TRIALID message to mark the trial start
    tk.sendMessage(f'TRIALID {trial}')

    # Start recording
    tk.startRecording(1, 1, 1, 1)

    # Assuming an image is presented in the task, and we would like
    # to have the same image in the background when visualizing data
    # in Data Viewer
    tk.sendMessage('!V IMGLOAD FILL woods.jpg')

    # Pretending that we are doing something for 2-sec
    pylink.pumpDelay(2000)

    # Stop recording
    tk.stopRecording()

    # Log a TRIAL_RESULT message to mark trial ends
    tk.sendMessage('TRIAL_RESULT 0')

# Wait for 100 to catch session end events
pylink.msecDelay(100)

# Close the EDF file and download it from the Host PC
tk.closeDataFile()
tk.receiveDataFile('imgload.edf', 'imgload_demo.edf')

# Close the link
tk.close()
```

The image will show up in Data Viewer when we load in the EDF data file (see Fig. 5.5). The IMGLOAD command is usually hidden in the Data panel. Ticking the "Include Display Command Messages" option in the Preference panel will show the IMGLOAD command along with other integration messages (e.g., TRIALID).

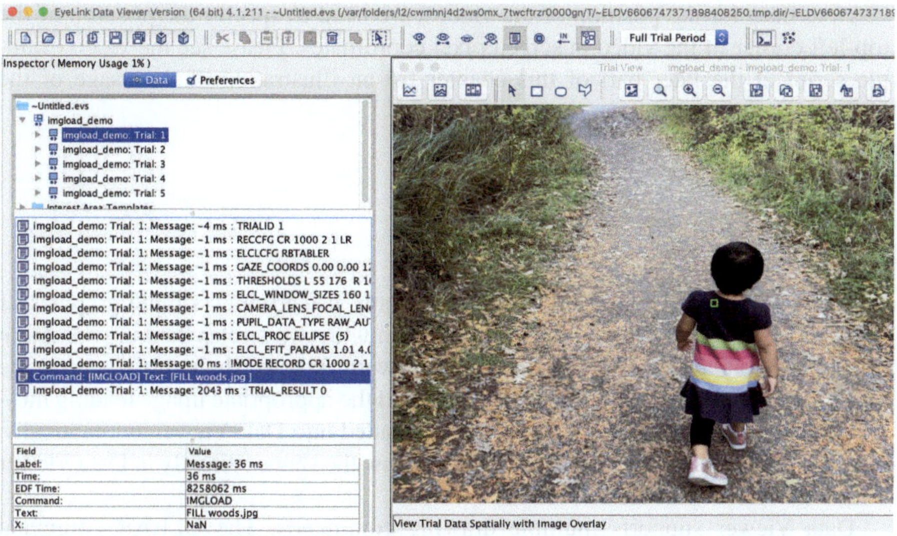

Fig. 5.5 A screenshot of the Data Viewer software showing the background image used for gaze data overlay

It is important to note that the timestamps of the background image message control when the image will show up during trial playback in Data Viewer. As such, these messages should be sent straight after the code that draws the image to the screen. If this is not possible, add a time offset to the IMGLOAD message so Data Viewer will correct its timing when loading in the EDF data file (see the Stroop script included in this chapter for an illustration of this approach).

Video

For tasks involving videos, it is desirable to load the video into Data Viewer to properly define dynamic interest areas and assign gaze data to the correct video frames. To use videos as background graphics when you playback the recorded data, Data Viewer needs to know the timing of each frame and the position of the video on the screen. For this purpose, we need to log "!V VFRAME" messages in the EDF data files for each frame of the video. The format of the VFRAME message is as follows.

```
!V VFRAME <frame_number> <pos_x> <pos_y> <movie_file>
```

Note that, for the position of the videos, we need to specify the position of the top-left corner of the video in screen pixel coordinates. Please see the video example script at the later part of this chapter for an illustration of the usage of the VFRAME messages.

Simple Drawing

Data Viewer also recognizes messages that request it to draw simple graphics as background for gaze overlay. Consider a screen with two circles flanking a fixation dot, and the one in red is the saccade target. Every time we show this screen to the participant, we could take a screenshot and send the appropriate image loading messages to the tracker so that the image can be loaded into Data Viewer automatically. An alternative and lightweight solution is to draw the shapes directly in Data Viewer with messages in predefined formats.

Data Viewer supports multiple drawing commands. The CLEAR command clears the screen; the DRAWLINE command draws a line; the DRAWBOX command draws a rectangle, whereas the FILLBOX command draws a filled box. The FIXPOINT command can be used to draw an empty or filled circle.

```
!V CLEAR <red> <green> <blue>
!V DRAWLINE <red> <green> <blue> <x_start> <y_start> <x_end> <y_end>
!V DRAWBOX <red> <green> <blue> <left> <top> <right> <bottom>
!V FILLBOX <red> <green> <blue> <left> <top> <right> <bottom>
!V FIXPOINT <target_red> <target_green> <target_blue> <erase_red> <erase_
green> <erase_blue> <x> <y> <outer_diameter> <inner_diameter>
```

We need to provide the color (in RGB format) and the screen pixel coordinates required for the drawing commands. For the FIXPOINT command, the target color (<target_*>) is the color for the outline, and the erase color (<erase_*>) is the color for the fixation center (or "fill" color). In the example script below, we use the FIXPOINT command to draw a black dot as the central fixation, an empty circle as the nontarget, and a filled red circle as the saccade target. Note that we also send the DISPLAY_COORDS message, so Data Viewer knows how large the screen is and where to draw the FIXPOINTs.

The screen reconstructed by Data Viewer from the messages may not precisely match that presented during testing, but it is sufficient for data visualization (see Fig. 5.6).

```
#!/usr/bin/env python
#
# Filename: simple_drawing.py
# Author: Zhiguo Wang
# Date: 2/6/2021
#
# Description:
# This script illustrates the various drawing commands
# supported by Data Viewer

import pylink


# Connect to the tracker
tk = pylink.EyeLink()

# Open an EDF on the Host; filename must not exceed 8 characters
tk.openDataFile('drawing.edf')

# Assume the screen resolution is 1280 x 800 pixels
SCN_W, SCN_H = (1280, 800)

# Pass the screen coordinates to the tracker
coords = f'screen_pixel_coords = 0 0 {SCN_W - 1} {SCN_H - 1}'
tk.sendCommand(coords)

# Record a DISPLAY_COORDS message to let Data Viewer know the
# correct screen resolution to use when visualizing the data
tk.sendMessage(f'DISPLAY_COORDS 0 0 {SCN_W - 1} {SCN_H - 1}')

# Run through five trials
for trial in range(1, 6):
    # Print out a message to show the current trial
    print(f'Trial #: {trial}')

    # Log a TRIALID message to mark trial start
    tk.sendMessage(f'TRIALID {trial}')

    # Start recording
    tk.startRecording(1, 1, 1, 1)

    # Clear the screen to show a white background
    tk.sendMessage('!V CLEAR 255 255 255')
    # Draw a central fixation dot
    tk.sendMessage('!V FIXPOINT 0 0 0 0 0 0 512 384 25 0')
```

```
        # Draw the non-target
        tk.sendMessage('!V FIXPOINT 0 0 0 255 255 255 312 384 80 75')
        # Draw the target
        tk.sendMessage('!V FIXPOINT 255 0 0 255 0 0 712 384 80 0')

        # Pretending that we are doing something for 2-sec
        pylink.pumpDelay(2000)

        # Stop recording
        tk.stopRecording()

        # Log a TRIAL_RESULT message to mark trial ends
        tk.sendMessage('TRIAL_RESULT 0')

    # Wait for 100 to catch session end events
    pylink.msecDelay(100)

    # Close the EDF file and download it from the Host PC
    tk.closeDataFile()
    tk.receiveDataFile('drawing.edf', 'drawing_demo.edf')

    # Close the link
    tk.close()
```

Draw List File

Suppose lots of image loading and drawing messages are needed to construct the stimulus screen in Data Viewer. In that case, a single message can be used to point to a single plain text file that contains all the Data Viewer integration commands. These files are known as draw list files, and they have the extension of ".dlf" (short for "draw list file"). The draw list file can contain simple drawing commands, as well as commands for loading background images. Similar to interest area set files, the commands in the draw list files do not require the "!V" prefix. In the example .dlf file below, the first command clears the screen and fills it with white color; the second command draws the image "color.bmp" with its top-left corner at screen pixel position (33, 33); the third command will draw a fixation point at (512, 384), on top of the background image.

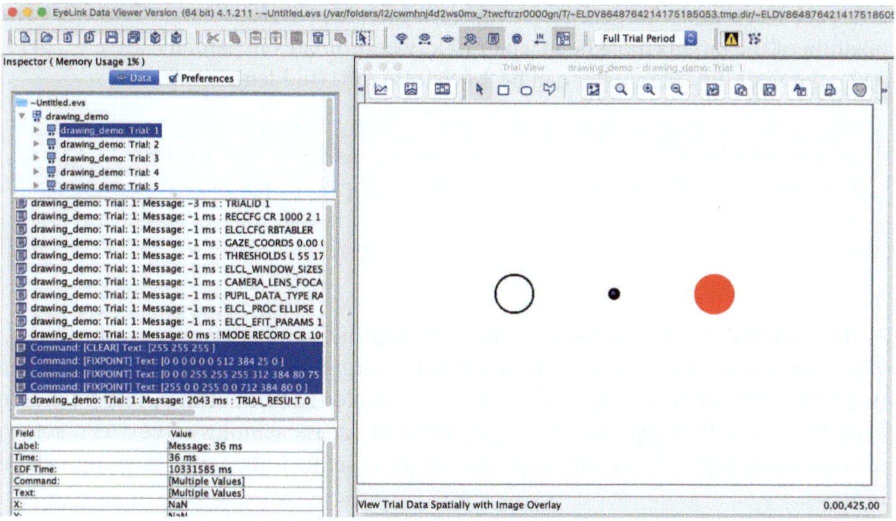

Fig. 5.6 Data Viewer reconstructs the background graphics from the drawing commands recorded in the EDF data files

```
CLEAR 255 255 255
IMGLOAD TOP_LEFT color.bmp 33 33 772 644
FIXPOINT 0 0 0 255 255 255 512 384 18 4
```

In the experimental script, we log a message in the following format to point to the draw list file. As with interest area set files, the <draw list file> parameter should contain the full path to the draw list file, relative to where the EDF data files are stored.

```
!V DRAW_LIST <draw list file>
```

Target Position

For experiments involving moving targets (i.e., smooth pursuit tasks), the target position can be logged with TARGET_POS messages, so that Data Viewer can use these messages to reconstruct the target movement trajectory. The format of these messages is presented below. A semicolon is required if there are two moving targets. Users can use any keyword to define the position of a single target; however, if you are defining two targets through the same command, please use the default

<Target1 Key> and <Target2 Key> values, which are "TARG1" and "TARG2." The position of the targets must be in a pair of x, y coordinates in parentheses, e.g., (411, 322); the visibility parameter can be 1 (visible) or 0 (hidden).

```
!V TARGET_POS <Target1 Key> <(target1 x, target1 y)> <target 1 visibility>
<target 1 step>; <Target2 Key> <(target2 x, target2 y)> <target 2 visibil-
ity> <target 2 step>
```

The TARGET_POS message is not updated until the next screen refresh, i.e., when the screen can change the position of the target. For the time interval between two TARGET_POS messages, the "step" parameter can be set to 0, in which case Data Viewer will interpolate the target location across samples in between screen refreshes (see Fig. 5.7). If the "step" parameter is set to 1, the target position will be in steps.

The target position data is needed to derive dependent measures like gain and phase lag in smooth pursuit tasks. We will present an example script in the following section to illustrate the use of the TARGET_POS message.

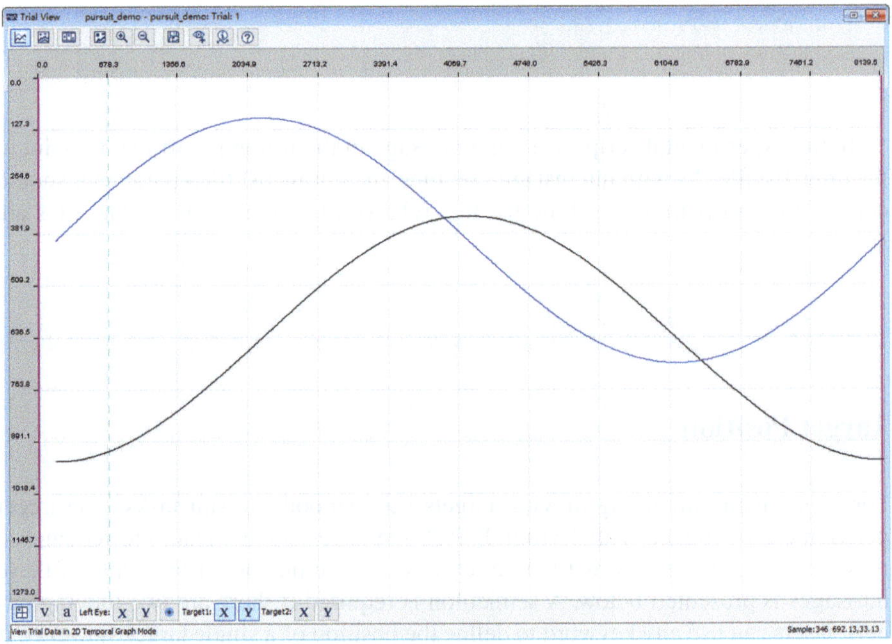

Fig. 5.7 Data Viewer reconstructed the target movement pattern from the TARGET_POS messages recorded in the EDF data files

Example Scripts in PsychoPy

This section will present three short scripts to illustrate the Data Viewer integration messages discussed in this chapter. These examples reflect my knowledge about the Pylink library and my personal coding preferences. SR Research has provided a set of Pylink examples that illustrates their recommended coding protocols. Those examples are included in the EyeLink Developer's Kit.

Stroop Task

The first example script implements the classic "Stroop task." As noted by Colin MacLeod in 1992, by that time, "it would be impossible to find anyone in cognitive psychology who does not have at least a passing acquaintance with the Stroop effect." MacLeod's claim still holds nearly three decades later. The Stroop task requires the participant to name the color in which a color word is printed. For example, the participant would say "blue" in response to the word "red" printed in blue. The semantic interference in these "incongruent" trials significantly delays the response, compared to "congruent" trials in which, for example, the word "green" is printed in green.

The PsychoPy script below illustrates the Stroop task and the Data Viewer integration messages, most notably the trial variable messages. The script is heavily commented on, but I will further explain some of the code.

```
#!/usr/bin/env python3
#
# Filename: Stroop_task.py
# Author: Zhiguo Wang
# Date: 2/6/2021
#
# Description:
# A Stroop task implemented in PsychoPy

import pylink
import os
import random
from psychopy import visual, core, event, monitors
from EyeLinkCoreGraphicsPsychoPy import EyeLinkCoreGraphicsPsychoPy

# Monitor resolution
SCN_W, SCN_H = (1280, 800)

# Step 1: Connect to the tracker
tk = pylink.EyeLink('100.1.1.1')
```

```
# Step 2: Open an EDF data file on the Host
tk.openDataFile('stroop.edf')
# Add preamble text (file header)
tk.sendCommand("add_file_preamble_text 'Stroop task demo'")

# Step 3: Set up tracking parameters
#
# Put the tracker in idle mode before we change its parameters
tk.setOfflineMode()

# Sample rate, 250, 500, 1000, or 2000 (depending on the tracker models,
# not all sample rate options are supported)
tk.sendCommand('sample_rate 500')

# Pass screen resolution  to the tracker
tk.sendCommand(f"screen_pixel_coords = 0 0 {SCN_W-1} {SCN_H-1}")

# Send a DISPLAY_COORDS message so Data Viewer knows the correct screen size
tk.sendMessage(f"DISPLAY_COORDS = 0 0 {SCN_W-1} {SCN_H-1}")

# Choose a calibration type, H3, HV3, HV5, HV13 (HV = horizontal/vertical)
tk.sendCommand("calibration_type = HV9")

# Step 4: # open a window for graphics and calibration
#
# Create a monitor object to store monitor information
customMon = monitors.Monitor('demoMon', width=35, distance=65)

# Open a PsychoPy window
win = visual.Window((SCN_W, SCN_H), fullscr=False,
                    monitor=customMon, units='pix')

# Request Pylink to use the PsychoPy window for calibration
graphics = EyeLinkCoreGraphicsPsychoPy(tk, win)
pylink.openGraphicsEx(graphics)

# Step 5: calibrate the tracker, and run through all the trials
calib_prompt = "Press LEFT to RED\nRIGHT to BLEU\n\nENTER to calibrate"
calib_msg = visual.TextStim(win, text=calib_prompt, color='white')
calib_msg.draw()
win.flip()

# Calibrate the tracker
tk.doTrackerSetup()

# Step 6: Run through all the trials
#
```

```
# Specify all possible experimental trials in a list; the columns are
# 'text', 'text_color', 'correct_answer' and "congruency"
my_trials = [
    ['red',   'red',  'left',  'cong'],
    ['red',   'blue', 'right', 'incg'],
    ['blue',  'blue', 'right', 'cong'],
    ['blue',  'red',  'left',  'incg']
    ]

# For convenience, define a run_trial function to group
# the lines of code repeatedly executed in each trial
def run_trial(params):
    """ Run a single trial

    params: a list containing tiral parameters, e.g.,
            ['red',   'red',   'left',  'cong']"""

    # Unpacking the parameters
    text, text_color, correct_answer, congruency = params

    # Prepare the stimuli
    word = visual.TextStim(win=win, text=text, font='Arial',
                           height=100.0, color=text_color)

    # Take the tracker offline
    tk.setOfflineMode()

    # Send a "TRIALID" message to mark the start of a trial
    tk.sendMessage(f"TRIALID {text} {text_color} {congruency}")

    # Record_status_message : show some info on the Host PC
    msg = f"record_status_message 'Congruency-{congruency}'"
    tk.sendCommand(msg)

    # Drift check/correction, params, x, y, draw_target, allow_setup
    tk.doDriftCorrect(int(SCN_W/2.0), int(SCN_H/2.0), 1, 1)

    # Put the tracker in idle mode before we start recording
    tk.setOfflineMode()

    # Start recording
    # params: file_sample, file_event, link_sampe, link_event (1-yes, 0-no)
    tk.startRecording(1, 1, 1, 1)

    # Wait for 100 ms to cache some samples
    pylink.msecDelay(100)
```

```
# Draw the target word on the screen
word.draw()
win.flip()
# Record the onset time of the stimuli
tar_onset = core.getTime()
# Send a message to mark the onset of visual stimuli
tk.sendMessage("stim_onset")

# Save a screenshot to use as background graphics in Data Viewer
if not os.path.exists('screenshots'):
    os.mkdir('screenshots')
screenshot = f'screenshots/cond_{text}_{text_color}.jpg'
win.getMovieFrame()
win.saveMovieFrames(screenshot)

# The command we used to take screenshots takes time to return
# we need to provide a "time offset" in the IMGLOAD message, so
# Data Viewer knows the correct onset time of the screen
msg_offset = int((core.getTime() - tar_onset) * 1000)
# Send an IMGLOAD message to let DV know which screenshot to load
scn_shot = '../' + screenshot
tk.sendMessage(f'{msg_offset} !V IMGLOAD FILL {scn_shot}')

# Clear bufferred events (in PsychoPy), then wait for key presses
event.clearEvents(eventType='keyboard')
gotKey = False
key_pressed, RT, ACC = ['None', 'None', 'None']
while not gotKey:
    keyp = event.getKeys(['left', 'right', 'escape'])
    if len(keyp) > 0:
        key_pressed = keyp[0]  # which key was pressed
        RT = core.getTime() - tar_onset  # response time
        # correct=1, incorrect=0
        ACC = int(key_pressed == correct_answer)

        # Send a message mark the key response
        tk.sendMessage(f"Key_resp {key_pressed}")
        gotKey = True

# Clear the window at the end of a trial
win.color = (0, 0, 0)
win.flip()

# Stop recording
tk.stopRecording()
```

```
    # Send trial variables to record in the EDF data file
    tk.sendMessage(f"!V TRIAL_VAR word {text}")
    tk.sendMessage(f"!V TRIAL_VAR color {text_color}")
    tk.sendMessage(f"!V TRIAL_VAR congruency {congruency}")
    pylink.pumpDelay(2)  # give the tracker a break
    tk.sendMessage(f"!V TRIAL_VAR key_pressed {key_pressed}")
    tk.sendMessage(f"!V TRIAL_VAR RT {round(RT * 1000)}")
    tk.sendMessage(f"!V TRIAL_VAR ACC {ACC}")

    # Send a 'TRIAL_RESULT' message to mark the end of the trial
    tk.sendMessage(f"TRIAL_RESULT {ACC}")

# Run a block of 8 trials, in random order
trials_to_test = my_trials[:]*2
random.shuffle(trials_to_test)
for trial in trials_to_test:
    run_trial(trial)

# Step 7: Close the EDF data file
tk.setOfflineMode()  # put the tracker in Offline
pylink.pumpDelay(100)  # wait for 100 ms
tk.closeDataFile()

# Step 8: Download EDF file to a local folder ('edfData')
msg = 'Downloading EDF file from the EyeLink Host PC ...'
edf = visual.TextStim(win, text=msg, color='white')
edf.draw()
win.flip()

if not os.path.exists('edfData'):
    os.mkdir('edfData')
tk.receiveDataFile('stroop.edf', 'edfData/stroop_demo.edf')

# Step 9: Close the connection to tracker, close graphics
tk.close()
win.close()
core.quit()
```

Custom Graphics for Calibration

At the beginning of the script, in addition to the various standard libraries (e.g., *pylink* and *random*) and the PsychoPy modules (*visual*, etc.), we also import a module named "EyeLinkCoreGraphicsPsychoPy." This module is included in the example scripts accompanying this book (available from https://github.com/zhiguo-eyelab/Pylink_book). A functionally more complete version of this module is included in the EyeLink Developer's Kit. We will discuss this module in detail in

Chap. 7. For now, it is sufficient to know that this module contains a set of functions for drawing the calibration target and the camera image, etc., on the Display PC during tracker calibration. To use this module, put it in the same folder as the experimental script, and then import it into the experimental script.

```
from EyeLinkCoreGraphicsPsychoPy import EyeLinkCoreGraphicsPsychoPy
```

Passing a tracker connection (*tk*) and a PsychoPy window (*win*) to *EyeLinkCoreGraphicsPsychoPy* will create a "graphics environment" (*graphics*). We then pass this graphics environment to *pylink.openGraphicsEx()* to request Pylink to use the custom graphics functions defined in the *EyeLinkCoreGraphicsPsychoPy* module.

```
graphics = EyeLinkCoreGraphicsPsychoPy(tk, win)
pylink.openGraphicsEx(graphics)
```

The DISPLAY_COORDS Message

As noted, it is necessary to log a "DISPLAY_COORDS" message in the EDF data file to store the resolution of the screen we used for data collection. Data Viewer needs this information to draw the background graphics and to overlay the gaze correctly.

```
# Send a DISPLAY_COORDS message so Data Viewer knows the correct screen size
tk.sendMessage(f"DISPLAY_COORDS = 0 0 {SCN_W-1} {SCN_H-1}")
```

The TRIALID and TRIAL_RESULT Messages

We log the TRIALID and TRIAL_RESULT messages to mark the start and end of a single trial. These messages must contain the keyword "TRIALID" or "TRIAL_RESULT," but they can also include additional information. In the example script above, we put the trial parameters (e.g., text, textColor, and congruency) in the TRIALID message. For the TRIAL_RESULT message, it includes the response accuracy variable (ACC). This additional information can be useful if you plan to analyze the data with tools other than Data Viewer.

```
# Send a "TRIALID" message to mark the start of a trial
    tk.sendMessage(f"TRIALID {text} {text_color} {congruency}")
    # Send a 'TRIAL_RESULT' message to mark the end of trial
    tk.sendMessage(f"TRIAL_RESULT {ACC}")
```

Record Status Message

As noted in the previous chapter, it can be helpful to show some information about the present trial on the Host PC, so the experimenter can easily monitor the progress of the task and ensure the participant is actively engaging in the task. In the example script above, the status message shows the word presented and its color, for instance, "word: blue, color: red" in the bottom-right corner of the Host PC screen.

```
# Record_status_message : show some info on the Host PC
msg = f"record_status_message 'Congruency-{congruency}'"
tk.sendCommand(msg)
```

Background Graphics: The IMGLOAD Message

The example script above shows the stimuli on the screen and then takes a screenshot and saves it to the hard drive. Saving screenshots to file takes a bit of time and may affect the timing of the IMGLOAD message. This potential issue can be avoided by adding a time offset (in milliseconds) to the beginning of the IMGLOAD message. In this example, the time offset for the IMGLOAD message was the difference between the current time (retrieved with *core.getTime())* and the time the image appeared on the screen (*tar_onset*).

```
msg_offset = int((core.getTime() - tar_onset) * 1000)
# Send an IMGLOAD message to let DV know which screenshot to load
scn_shot = '../' + screenshot
tk.sendMessage(f'{msg_offset} !V IMGLOAD FILL {scn_shot}')
```

The lines of code for taking screenshots are included in the script to illustrate the "time offset" we can include in the IMGLOAD message. Because the script will generate only four unique screenshots (e.g., "cond_red_red.jpg"), you may use an if statement to check if the screenshot for the current trial is already in the "screenshots" folder. If so, skip the lines of code for capturing screenshots to get rid of the message time offset.

Trial Variable Messages

For this Stroop task, we store the parameters of each trial in a list. There are two target words ("red" and "blue"), and each could be in red or blue, so we have a total of four possible combinations. The four items specified in the list correspond to the word presented, the word color, and the correct keyboard response (left or right arrow key). An additional variable (congruency) is included in the list to label the condition the current experimental trial belongs to (i.e., congruent or incongruent).

```
# Specify all possible experimental trials in a list; the columns are
# 'text', 'text_color', 'correct_answer' and "congruency"
my_trials = [
    ['red',   'red',  'left',  'cong'],
    ['red',   'blue', 'right', 'incg'],
    ['blue',  'blue', 'right', 'cong'],
    ['blue',  'red',  'left',  'incg']
    ]
```

The parameters of each trial, along with the response data (e.g., reaction time and accuracy), are saved in the EDF data files to facilitate offline data analysis. The example script above includes several trial variable messages. These variables will show up in the Trial Variable Manager when the EDF data file is loaded into Data Viewer. The EyeLink Host PC can handle multiple messages that arrive at roughly the same time; nevertheless, it is recommended to call *pylink.pumpDelay()* to give the tracker a break (of a couple of milliseconds) every three to four messages, if you have a lot of messages to send.

```
    # Send trial variables to record in the EDF data file
    tk.sendMessage(f"!V TRIAL_VAR word {text}")
    tk.sendMessage(f"!V TRIAL_VAR color {text_color}")
    tk.sendMessage(f"!V TRIAL_VAR congruency {congruency}")
    pylink.pumpDelay(2)  # give the tracker a break
    tk.sendMessage(f"!V TRIAL_VAR key_pressed {key_pressed}")
    tk.sendMessage(f"!V TRIAL_VAR RT {round(RT * 1000)}")
    tk.sendMessage(f"!V TRIAL_VAR ACC {ACC}")
```

Messages Marking Critical Trial Events

To properly analyze and visualize eye movement data in Data Viewer, in addition to the Data Viewer integration messages described above, it is also important to have messages that mark the occurrence of critical trial events. Without these messages, it would be impossible to tell what events occurred at what time during the recording. On a practical level, these messages (e.g., "stim_onset," "key_resp") allow you to define time windows (i.e., "interest period" in Data Viewer terminology) that are critical to your analysis protocol.

```
    # send a message to mark the onset of visual stimuli
    tk.sendMessage("stim_onset")
```

In the example script above, we send a message to the tracker when the visual stimuli appear on the screen. When the participant issues a keyboard response, we log this event with the following message.

```
            # Send a message mark the key response
            tk.sendMessage(f"Key_resp {key_pressed}")
```

Video

Support for video playback in PsychoPy 3 has improved. To properly analyze eye movements recorded during video viewing, it is important to play the video in Data Viewer so dynamic interest areas can be defined if needed. The video example presented here shows how to embed Data Viewer integration messages in a PsychoPy script.

```
#!/usr/bin/env python3
#
# Filename: video_task.py
# Author: Zhiguo Wang
# Date: 2/6/2021
#
# Description:
# Play video and record eye movements in Psychopy

import pylink
import os
import random
from psychopy import visual, core, event, monitors
from EyeLinkCoreGraphicsPsychoPy import EyeLinkCoreGraphicsPsychoPy
from psychopy.constants import FINISHED

# Screen resolution
SCN_W, SCN_H = (1280, 800)

# SETP 1: Connect to the tracker
tk = pylink.EyeLink('100.1.1.1')

# Step 2: Open an EDF data file on the Host
tk.openDataFile('video.edf')
# Add preamble text (file header)
tk.sendCommand("add_file_preamble_text 'Movie playback demo'")

# Step 3: Set up tracking parameters
#
# put the tracker in idle mode before we change its parameters
```

```
tk.setOfflineMode()

# Sample rate, 250, 500, 1000, or 2000 (depending on the tracker models,
# not all sample rate options are supported)
tk.sendCommand('sample_rate 500')

# Pass screen resolution  to the tracker
tk.sendCommand(f"screen_pixel_coords = 0 0 {SCN_W-1} {SCN_H-1}")

# Send a DISPLAY_COORDS message so Data Viewer knows the correct screen size
tk.sendMessage(f"DISPLAY_COORDS = 0 0 {SCN_W-1} {SCN_H-1}")

# Choose a calibration type, H3, HV3, HV5, HV13 (HV = horizontal/vertical)
tk.sendCommand("calibration_type = HV9")

# Step 4: # open a window for graphics and calibration
#
# Create a monitor object to store monitor information
customMon = monitors.Monitor('demoMon', width=35, distance=65)

# Open a PsychoPy window
win = visual.Window((SCN_W, SCN_H), fullscr=False,
                    monitor=customMon, units='pix')

# Request Pylink to use the PsychoPy window for calibration
graphics = EyeLinkCoreGraphicsPsychoPy(tk, win)
pylink.openGraphicsEx(graphics)

# Step 5: Calibrate the tracker, and run through all the trials
calib_prompt = "Press ENTER to calibrate the tracker"
calib_msg = visual.TextStim(win, text=calib_prompt, color='white', )
calib_msg.draw()
win.flip()

# Calibrate the tracker
tk.doTrackerSetup()

# Step 6: Run through a couple of trials
# put the videos we would like to play in a list
trials = [
    ['t1', 'driving.mov'],
    ['t2', 'driving.mov']
    ]

# Here, we define a helper function to group the code executed on each trial
def run_trial(pars):
    """ pars corresponds to a row in the trial list"""
```

```
    # Retrieve parameters from the trial list
    trial_num, movie_file = pars

    # Load the video to display
    mov = visual.MovieStim3(win, filename=movie_file, size=(960, 540))

    # Take the tracker offline
    tk.setOfflineMode()

    # Send the standard "TRIALID" message to mark the start of a trial
    tk.sendMessage(f"TRIALID {trial_num} {movie_file}")

    # Record_status_message : show some info on the Host PC
    msg = f"record_status_message 'Movie File: {movie_file}'"
    tk.sendCommand(msg)

    # Drift check/correction, params, x, y, draw_target, allow_setup
    tk.doDriftCorrect(int(SCN_W/2.0), int(SCN_H/2.0), 1, 1)

    # Put the tracker in idle mode before we start recording
    tk.setOfflineMode()

    # Start recording
    # params: file_sample, file_event, link_sampe, link_event (1-yes, 0-no)
    tk.startRecording(1, 1, 1, 1)

    # Wait for 100 ms to cache some samples
    pylink.msecDelay(100)

    # The size of the video
    mo_width, mo_height = mov.size

    # play the video till the end
    frame_n = 0
    prev_frame_timestamp = mov.getCurrentFrameTime()
    while mov.status is not FINISHED:
        # draw a movie frame and flip the video buffer
        mov.draw()
        win.flip()

        # if a new frame is drawn, check frame timestamp and
        # send a VFRAME message
        current_frame_timestamp = mov.getCurrentFrameTime()
        if current_frame_timestamp != prev_frame_timestamp:
            frame_n += 1
            # send a message to mark the onset of each video frame
            tk.sendMessage(f'Video_Frame: {frame_n}')
```

```
            # VFRAME message: "!V VFRAME frame_num movie_pos_x,
            # movie_pos_y, path_to_movie_file"
            x = int(SCN_W/2.0 - mo_width/2.0)
            y = int(SCN_H/2.0 - mo_height/2.0)
            path_to_movie = '../' + movie_file
            msg = f"!V VFRAME {frame_n} {x} {y} {path_to_movie}"
            tk.sendMessage(msg)
            prev_frame_timestamp = current_frame_timestamp

    # Send a message to mark the end of the video
    tk.sendMessage("Video_terminates")

    # Clear the subject display
    win.color = (0, 0, 0)
    win.flip()

    # Stop recording
    tk.stopRecording()

    # Send a'TRIAL_RESULT' message to mark the end of trial
    tk.sendMessage('TRIAL_RESULT')

# Run a block of 2 trials, in random order
test_list = trials[:]
random.shuffle(test_list)
for trial in test_list:
    run_trial(trial)

# Step 7: Close the EDF data file and put the tracker in idle mode
tk.setOfflineMode()  # put the tracker in Offline
pylink.pumpDelay(100)  # wait for 100 ms
tk.closeDataFile()

# Step 8: Download EDF file to a local folder ('edfData')
msg = 'Downloading EDF file from the EyeLink Host PC ...'
edfTransfer = visual.TextStim(win, text=msg, color='white')
edfTransfer.draw()
win.flip()

if not os.path.exists('edfData'):
    os.mkdir('edfData')
tk.receiveDataFile('video.edf', 'edfData/video_demo.edf')

# Step 9: Close the connection to tracker, close graphics
tk.close()
win.close()
core.quit()
```

This script is very similar to the Stroop example. The only lines of code worth mentioning are those related to video playback. First, we load in the video clip to create a *visual.Movie3()* object; we then repeatedly call the *draw()* method and flip the video buffers in a *while* loop to show the video frames on the screen.

```
# Load the video to display
mov = visual.MovieStim3(win, filename=movie_file, size=(960, 540))

while mov.status is not FINISHED:
    # draw a movie frame and flip the video buffer
    mov.draw()
    win.flip()
```

For Data Viewer to correctly play a video, appropriate messages need to be written to the EDF file to let Data Viewer know when to load each video frame. The message starts with the "!V VFRAME" command, followed by the current frame number, the x and y pixel coordinates of the top-left corner of the video, and the path to the video file, relative to the folder storing the EDF data file.

```
!V VFRAME frame_num movie_pos_x, movie_pos_y, path_to_file
```

We can use the *getCurrentFrameTime()* function to retrieve the current frame time relative to the start of a video file. Before the start of video playback, the *getCurrentFrameTime()* function will return a negative timestamp. The script stores this negative timestamp in a variable named *prev_frame_timestamp*. In the while loop, we check if the current frame time is different from the *prev_frame_timestamp* variable; if we know a new frame is on the screen, we send a standard message to mark the onset of this frame. Then, we record a separate VFRAME message so Data Viewer can use it to draw the video frames during playback. At the end of the while loop, of course, we need to assign the current frame time to the *prev_frame_timestamp* variable, so we can use it to check if a new frame is on the screen in the next iteration.

```
# play the video till the end
frame_n = 0
prev_frame_timestamp = mov.getCurrentFrameTime()
while mov.status is not FINISHED:
    # draw a movie frame and flip the video buffer
    mov.draw()
    win.flip()

    # if a new frame is drawn, check frame timestamp and
    # send a VFRAME message
```

```
        current_frame_timestamp = mov.getCurrentFrameTime()
        if current_frame_timestamp != prev_frame_timestamp:
            frame_n += 1
            # send a message to mark the onset of each video frame
            tk.sendMessage(f'Video_Frame: {frame_n}')
            # VFRAME message: "!V VFRAME frame_num movie_pos_x,
            # movie_pos_y, path_to_movie_file"
            x = int(SCN_W/2 - mo_width/2)
            y = int(SCN_H/2 - mo_height/2)
            path_to_movie = '../' + movie_file
            msg = f"!V VFRAME {frame_n} {x} {y} {path_to_movie}"
            tk.sendMessage(msg)
            prev_frame_timestamp = current_frame_timestamp
```

Pursuit Task

The smooth pursuit task is frequently used in clinical studies. In the task, the participant follows a visual target that moves in a pattern. In the example script below, the visual target follows a predefined circular movement pattern. The code relevant to eye tracking is mostly the same as in the previous example scripts. We will discuss the lines of code that are specific to this pursuit task in detail.

```
#!/usr/bin/env python3
#
# Filename: pursuit_task.py
# Author: Zhiguo Wang
# Date: 2/6/2021
#
# Description:
# A simple smooth pursuit task implemented in PsychoPy

import pylink
import os
import random
from psychopy import visual, core, event, monitors
from EyeLinkCoreGraphicsPsychoPy import EyeLinkCoreGraphicsPsychoPy
from math import sin, pi

# Monitor resolution
SCN_W, SCN_H = (1280, 800)

# Step 1: Connect to the tracker
tk = pylink.EyeLink('100.1.1.1')
```

```
# Step 2: Open an EDF data file on the Host
tk.openDataFile('pursuit.edf')
# Add preamble text (file header)
tk.sendCommand("add_file_preamble_text 'Smooth pursuit demo'")

# Step 3: Setup Host parameters
# put the tracker in idle mode before we change its parameters
tk.setOfflineMode()
pylink.msecDelay(50)

# Sample rate, 250, 500, 1000, or 2000 (depending on the tracker models,
# not all sample rate options are supported)
tk.sendCommand('sample_rate 500')

# Pass screen resolution  to the tracker
tk.sendCommand(f"screen_pixel_coords = 0 0 {SCN_W-1} {SCN_H-1}")

# Send a DISPLAY_COORDS message so Data Viewer knows the correct screen size
tk.sendMessage(f"DISPLAY_COORDS = 0 0 {SCN_W-1} {SCN_H-1}")

# Choose a calibration type, H3, HV3, HV5, HV13 (HV = horizontal/vertical)
tk.sendCommand("calibration_type = HV9")

# Step 4: # open a window for graphics and calibration
#
# Create a monitor object to store monitor information
customMon = monitors.Monitor('demoMon', width=35, distance=65)

# Open a PsychoPy window
win = visual.Window((SCN_W, SCN_H), fullscr=True,
                    monitor=customMon, units='pix')

# Request Pylink to use the PsychoPy window for calibration
graphics = EyeLinkCoreGraphicsPsychoPy(tk, win)
pylink.openGraphicsEx(graphics)

# Step 5: prepare the pursuit target, the clock and the movement parameters
target = visual.GratingStim(win, tex=None, mask='circle', size=25)
pursuitClock = core.Clock()

# Parameters for the Sinusoidal movement pattern
# [amp_x, amp_y, phase_x, phase_y, freq_x, freq_y]
mov_pars = [
    [300, 300, pi*3/2, 0, 1/8.0, 1/8.0],
    [300, 300, pi/2, 0, 1/8.0, 1/8.0]
    ]
```

```
# Step 6: Provide instructions and calibrate the tracker
calib_prompt = 'Press Enter twice to calibrate the tracker'
calib_msg = visual.TextStim(win, text=calib_prompt, color='white',
units='pix')
calib_msg.draw()
win.flip()

# Calibrate the tracker
tk.doTrackerSetup()

# Step 7: Run through a couple of trials
# define a function to group the code that will executed on each trial
def run_trial(trial_duration, movement_pars):
    """ Run a smooth pursuit trial

    trial_duration: the duration of the pursuit movement
    movement_pars: [amp_x, amp_y, phase_x, phase_y, freq_x, freq_y]
    The following equation defines a sinusoidal movement pattern
    y(t) = amplitude * sin(2 * pi * frequency * t + phase)
    for circular or elliptic movements, the phase in x and y directions
    should be pi/2 (direction matters)."""

    # Parse the movement pattern parameters
    amp_x, amp_y, phase_x, phase_y, freq_x, freq_y = movement_pars

    # Take the tracker offline
    tk.setOfflineMode()

    # Send the standard "TRIALID" message to mark the start of a trial
    tk.sendMessage("TRIALID")

    # Record_status_message : show some info on the Host PC
    tk.sendCommand("record_status_message 'Pursuit demo'")

    # Drift check/correction, params, x, y, draw_target, allow_setup
    tar_x = amp_x*sin(phase_x)
    tar_y = amp_y*sin(phase_y)
    target.pos = (tar_x, tar_y)
    target.draw()
    win.flip()

    tk.doDriftCorrect(int(tar_x + SCN_W/2.0), int(SCN_H/2.0 - tar_y), 0,1)
    # Put the tracker in idle mode before we start recording
    tk.setOfflineMode()

    # Start recording
    # params: file_sample, file_event, link_sampe, link_event (1-yes, 0-no)
```

```
tk.startRecording(1, 1, 1, 1)

# Wait for 100 ms to cache some samples
pylink.msecDelay(100)

# Send a message to mark movement onset
frame = 0
while True:
    target.pos = (tar_x, tar_y)
    target.draw()
    win.flip()
    flip_time = core.getTime()
    frame += 1
    if frame == 1:
        tk.sendMessage('Movement_onset')
        move_start = core.getTime()
    else:
        _x = int(tar_x + SCN_W/2.0)
        _y = int(SCN_H/2.0 - tar_y)
        tar_msg = f'!V TARGET_POS target {_x}, {_y} 1 0'
        tk.sendMessage(tar_msg)

    time_elapsed = flip_time - move_start

    # update the target position
    tar_x = amp_x * sin(2 * pi * freq_x * time_elapsed + phase_x)
    tar_y = amp_y * sin(2 * pi * freq_y * time_elapsed + phase_y)

    # break if the time elapsed exceeds the trial duration
    if time_elapsed > trial_duration:
        break

# clear the window
win.color = (0, 0, 0)
win.flip()

# Stop recording
tk.stopRecording()

# Send trial variables to record in the EDF data file
tk.sendMessage(f"!V TRIAL_VAR amp_x {amp_x:.2f}")
tk.sendMessage(f"!V TRIAL_VAR amp_y {amp_y:.2f}")
tk.sendMessage(f"!V TRIAL_VAR phase_x {phase_x:.2f}")
pylink.pumpDelay(2)  # give the tracker a break
tk.sendMessage(f"!V TRIAL_VAR phase_y {phase_y:.2f}")
tk.sendMessage(f"!V TRIAL_VAR freq_x {freq_x:.2f}")
```

```
        tk.sendMessage(f"!V TRIAL_VAR freq_y {freq_y:.2f}")
        tk.sendMessage(f"!V TRIAL_VAR duration {trial_duration:.2f}")

        # Send a 'TRIAL_RESULT' message to mark the end of trial
        tk.sendMessage('TRIAL_RESULT')

# Run a block of 2 trials, in random order
test_list = mov_pars[:]
random.shuffle(test_list)
for trial in test_list:
    run_trial(8.0, trial)
# Step 8: Close the EDF data file and put the tracker in idle mode
tk.setOfflineMode()  # put the tracker in Offline
pylink.pumpDelay(100)  # wait for 100 ms
tk.closeDataFile()
# Step 9: Downlad EDF file to a local folder ('edfData')
msg = 'Downloading EDF file from the EyeLink Host PC ...'
edf = visual.TextStim(win, text=msg, color='white')
edf.draw()
win.flip()

if not os.path.exists('edfData'):
    os.mkdir('edfData')
tk.receiveDataFile('pursuit.edf', 'edfData/pursuit_demo.edf')

# Step 10: Close the connection to tracker, close graphics
tk.close()
win.close()
core.quit()
```

Researchers frequently use a sinusoidal movement pattern when implementing smooth pursuit tasks. The x, y position of the visual target follows the following equations.

```
tar_x = amplitude_x * sin(2 * pi * freq_x * time_elapsed + phase_x)
tar_y = amplitude_y * sin(2 * pi * freq_y * time_elapsed + phase_y)
```

In the equations, the *amplitude_** parameters determine the amplitude of the sine wave. For instance, with an amplitude of 300 pixels in the horizontal direction, the target will move in the −300 to the +300-pixel range, assuming 0 is the center of the screen. The *frequency_** parameters specify the frequency of the sine wave—the numbers of oscillations (cycles) that occur in a second. The *t* parameter specifies how much time has elapsed since movement onset. The *phase_** parameters determine the start phase of the movement, i.e., where to start the movement. For instance, in the horizontal direction, the movement starts (at $t = 0$) from the center if the start phase is 0 [sin(0) = 0], from the right if the start phase is pi/2 [sin(pi/2) = 1], and from the left if the start phase is pi*3/2 [sin(pi*3/2) = −1].

By manipulating these parameters, we can flexibly configure the target movement pattern, velocity, and amplitude. For vertical movement, set the amplitude in the horizontal direction to 0 (*amplitude_x = 0*); for horizontal movement, set the amplitude in the vertical direction to 0 (*amplitude_y = 0*). The start phase in the horizontal and vertical directions should be pi/2 apart for circular or elliptic movement. The movement will be clockwise if the start phase in the vertical direction is pi/2 greater than the horizontal direction; the movement will be counterclockwise if the start phase is pi/2 greater in the vertical direction. Other start phase differences between the horizontal and vertical directions will create a complex Lissajous trajectory.

We need to send the following message to log the target position in the EDF data file. Note that the origin (0, 0) of the screen coordinates used in PsychoPy is the screen center. We need to convert the target position, so it refers to the top-left corner of the screen as (0, 0), the origin assumed by Data Viewer. With the help of a *while* loop, the TARGET_POS message is sent to the tracker every time the screen redraws.

```
_x = int(tar_x + SCN_W/2.0)
_y = int(SCN_H/2.0 - tar_y)
tar_msg = f'!V TARGET_POS target {_x}, {_y} 1 0'
tk.sendMessage(tar_msg)
```

This chapter discusses the various messages to log into the EDF data files to prepare for analysis and visualization in the Data Viewer software. We will discuss alternative ways to analyze and visualize eye movement data in Chapter 8. Still, as noted, Data Viewer is a very convenient tool for a wide range of eye-tracking analysis tasks. I would encourage you to log the messages discussed in this chapter, even if you have no plan to use Data Viewer for analysis and visualization. These messages are part of a mature data logging protocol, which can be extremely useful when analyzing your data in MATLAB, Python, or any other programming languages or tools.

Chapter 6
Accessing Gaze Data During Recording

Contents

The eye-tracking applications discussed in Chaps. 4 and 5 involve "passive recording" only. We record eye movements while the participant is performing a task, but we do not use the eye movement data in any way during the task. In many research scenarios, however, it is necessary to access real-time gaze data to drive the presentation of the stimuli. For instance, in a moving window task, the observer can view a small screen region around the current gaze position. In a saccadic adaptation task, the target is shifted immediately following saccade onset. This chapter will focus on the retrieval of eye movement data during recording.

Samples and Events

Two types of data are available from the EyeLink Host PC during recording—samples and events. A detailed discussion of these two types of data is presented in Chap. 8. For now, it is sufficient to know that samples are the "raw" eye movement data and comprise the x, y gaze position and pupil size. They are provided every 1 or 2 ms, depending on the sampling rate. Events are fixations, saccades, and blinks detected from the sample data.

The parsing of sample data into fixations and saccades can be a complicated topic, and different research fields have different conventions and approaches. The algorithm used by the EyeLink trackers is velocity-based and is applied by the Host

Z. Wang, *Eye-Tracking with Python and Pylink*,
https://doi.org/10.1007/978-3-030-82635-2_6

PC software during recording. A saccade is detected if the eye velocity or acceleration exceeds predefined thresholds, and the same saccade terminates when both velocity and acceleration fall below thresholds. The onset of a new fixation follows the termination of the leading saccade; the start of a saccade is preceded by the end of the preceding fixation (see Fig. 6.1). You can find further details of the saccade parsing algorithm in the EyeLink user manuals.

The high sampling rate of the EyeLink trackers allows the identification of eye events during recording, and the tracker can make these events available over the link. Sample data can be accessed in a near real-time fashion (see Fig. 6.2), with a delay in the one- to two-sample range (depending on the sampling rate and filter setting). On the other hand, eye events require some time to identify and cannot be accessed in a real-time manner. For instance, for detecting saccade onset, the algorithm includes a brief verification period to ensure velocity or acceleration is indeed above the thresholds. Consequently, event data is typically delayed by about 20–40 ms over the link, depending on the tracker configuration.

The EyeLink trackers allow flexible control of the eye movement data. When you call *startRecording()*, it is mandatory to specify four parameters to let the tracker know whether sample and event data should be saved to file and made available over the link.

<file_samples> -- 1, writes samples to EDF; 0, disables sample recording
<file_events> -- 1, writes events to EDF; 0, disables event recording

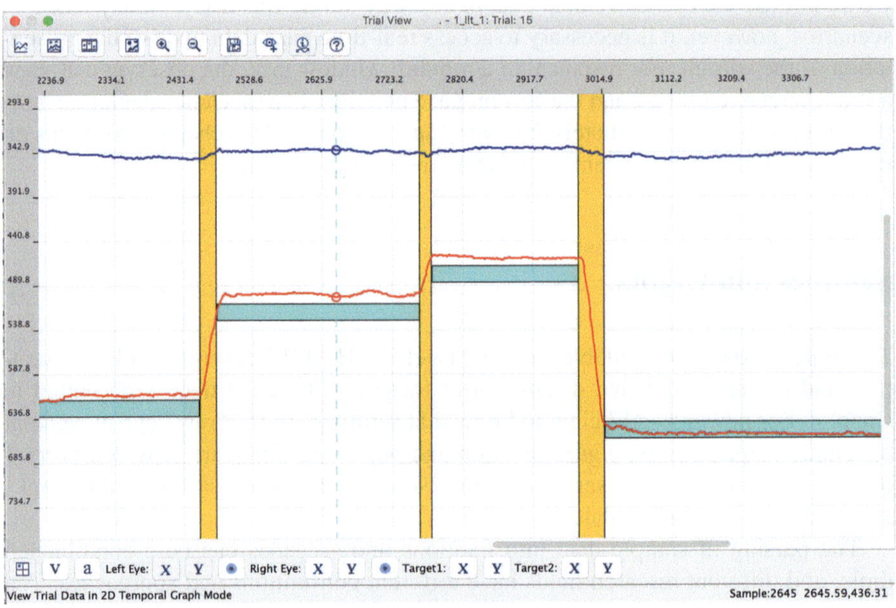

Fig. 6.1 The EyeLink tracker detects eye events based on instantaneous velocity and acceleration. In this temporal graph, the X, Y gaze traces are plotted in blue and red. Blue boxes represent fixations, and vertical stripes represent saccades

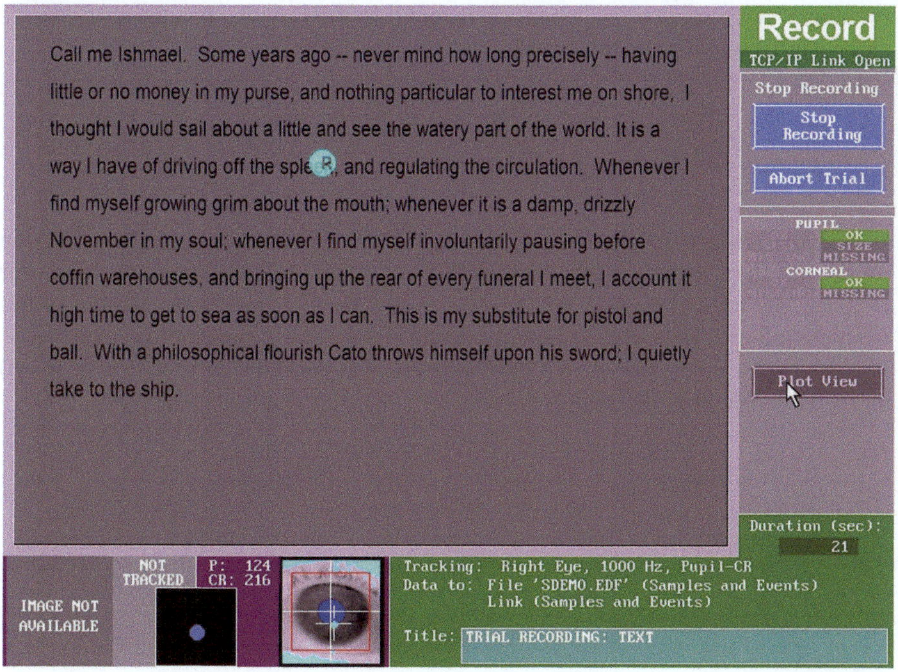

Fig. 6.2 A screenshot of the EyeLink Host PC showing that real-time gaze data is used to drive a gaze cursor. The eye movement data is also accessible from another device through an Ethernet link to the tracker

<link_samples> -- 1, sends samples through link; 0, disables link sample
<link_events> -- 1, sends events through link; 0, disables link event

The following command sets the tracker to record both events and samples in the EDF data file and make both events and samples available over the link. Switching the *<link_samples>* and *<link_events>* parameters to 0 will disable sample and event access over the link during recording.

```
tk.startRecording(1, 1, 1, 1)
```

Accessing Sample Data

Once we have started recording and made sure that the eye movement data will be available over the link, we can use various commands to retrieve samples or events during testing. The retrieval of samples is helpful in gaze-contingent tasks, for instance, the classic moving window task by McConkie and Rayner (1975; see Rayner, 2014, for a review). Fast online retrieval of gaze samples is also necessary for boundary-crossing tasks, where we manipulate the stimuli when the gaze crosses an invisible boundary on the screen.

Commands for Accessing Samples

We use the *getNewestSample()* function to retrieve the latest sample. This function will return the latest *sample*, which has lots of properties that we can extract. For instance, the following code snippet determines whether the newest sample comes from the left eye, right eye, or both eyes, and it also returns the resolution data (i.e., how many pixels correspond to 1 degree of visual angle at the current gaze position) and the timestamp of the sample.

```
smp = tk.getNewestSample() # retrieve the latest sample
is_left = smp.isLeftSample() # left eye sample
is_right = smp.isRightSample() # right eye sample
is_bino = smp.isBinocular() # binocular recording is on
res = smp.getPPD() # 1 deg = how many pixels at the current gaze position
time_stamp = smp.getTime() # timestamp on the tracker
```

To get the current gaze position, we call *getLeftEye()* or *getRightEye()* to get the actual *SampleData* for the left and right eye, respectively. In the following code snippet, if the right eye data is available, we retrieve the gaze position from the right eye data stream; otherwise, we extract data from the left eye data stream.

```
# Poll the latest samples
dt = smp.getNewestSample()
if smp is not None:
    # Grab gaze, HREF, raw, & pupil size data
    if smp.isRightSample():
        gaze = smp.getRightEye().getGaze()
        href = smp.getRightEye().getHREF()
        raw = smp.getRightEye().getRawPupil()
        pupil = smp.getRightEye().getPupilSize()
    elif smp.isLeftSample():
        gaze = smp.getLeftEye().getGaze()
        href = smp.getLeftEye().getHREF()
        raw = smp.getLeftEye().getRawPupil()
        pupil = smp.getLeftEye().getPupilSize()
```

In addition to gaze position (in screen pixel coordinates), the samples may also contain HREF or RAW data, depending on the tracker configuration. HREF is head-referenced data, i.e., eye rotations relative to the head; it can be helpful in physiological studies examining the actual movement of the eye. RAW data is the raw data from the camera sensor; it is useful in studies where researchers would like to calibrate the tracker with their calibration routines. For instance, one can adjust signal gain and offset to map the RAW signal range to a known screen region. The HREF data can be retrieved with *getHREF()*, RAW data can be retrieved with *getRawPupil()*, and pupil size data can be retrieved with *getPupilSize()*.

Below is a short script to illustrate how to retrieve the various types of sample data, including gaze, HREF, RAW, and pupil size data.

```
#!/usr/bin/env python3
#
# Filename: sample_retrieval.py
# Author: Zhiguo Wang
# Date: 2/6/2021
#
# Description:
# A short script illustrating online retrieval of sample data

import pylink


# Connect to the tracker
tk = pylink.EyeLink('100.1.1.1')

# Open an EDF data file on the Host PC
tk.openDataFile('smp_test.edf')

# Put the tracker in offline mode before we change tracking parameters
tk.setOfflineMode()

# Set sample rate to 1000 Hz
tk.sendCommand('sample_rate 1000')

# Make gaze, HREF, and raw (PUPIL) data available over the link
sample_flag = 'LEFT,RIGHT,GAZE,GAZERES,PUPIL,HREF,AREA,STATUS,INPUT'
tk.sendCommand(f'link_sample_data = {sample_flag}')

# Open an SDL window for calibration
pylink.openGraphics()
```

```
# Set up the camera and calibrate the tracker
tk.doTrackerSetup()

# Put tracker in idle/offline mode before we start recording
tk.setOfflineMode()

# Start recording
error = tk.startRecording(1, 1, 1, 1)

# Wait for a moment
pylink.msecDelay(100)

# Open a plain text file to store the retrieved sample data
text_file = open('sample_data.csv', 'w')

# Current tracker time
t_start = tk.trackerTime()
smp_time = -1
while True:
    # Break after 5 seconds have elapsed
    if tk.trackerTime() - t_start > 5000:
        break

    # Poll the latest samples
    smp = tk.getNewestSample()
    if smp is not None:
        # Grab gaze, HREF, raw, & pupil size data
        if smp.isRightSample():
            gaze = smp.getRightEye().getGaze()
            href = smp.getRightEye().getHREF()
            raw = smp.getRightEye().getRawPupil()
            pupil = smp.getRightEye().getPupilSize()
        elif smp.isLeftSample():
            gaze = smp.getLeftEye().getGaze()
            href = smp.getLeftEye().getHREF()
            raw = smp.getLeftEye().getRawPupil()
            pupil = smp.getLeftEye().getPupilSize()

        timestamp = smp.getTime()

        # Save gaze, HREF, raw, & pupil data to the plain text
        # file, if the sample is new
        if timestamp > smp_time:
            smp_data = map(str, [timestamp, gaze, href, raw, pupil])
            text_file.write('\t'.join(smp_data) + '\n')
```

```
            smp_time = timestamp

# Stop recording
tk.stopRecording()

# Close the plain text file
text_file.close()
# Put the tracker to the offline mode
tk.setOfflineMode()
# Close the EDF data file on the Host
tk.closeDataFile()

# Download the EDF data file from Host
tk.receiveDataFile('smp_test.edf', 'smp_test.edf')

# Close the link to the tracker
tk.close()

# Close the window
pylink.closeGraphics()
```

In the script, we first connect to the tracker, open a calibration window, calibrate the tracker, and close the calibration window. We then send two commands to the tracker to ensure the tracker operates at 1000 Hz and, most importantly, make the HREF and raw PUPIL data available over the link. The second command (*link_sample_data*) controls what types of sample data can be accessed over the link during recording. There is no need to delve into the details here; please see the EyeLink user manuals for more information about the sample data flags (e.g., GAZE, HREF, PUPIL) that can be included in this command.

```
# Set sample rate to 1000 Hz
tk.sendCommand('sample_rate 1000')

# Make gaze, HREF, and raw (PUPIL) data available over the link
sample_flag = 'LEFT,RIGHT,GAZE,GAZERES,PUPIL,HREF,AREA,STATUS,INPUT'
tk.sendCommand(f'link_sample_data = {sample_flag}')
```

For an illustration, we open a plain text file to store the sample data we accessed during recording. Note that, before we enter the *while* loop, *trackerTime()* is called to get the current time on the EyeLink Host PC; in the *while* loop, we repeatedly call this function to check if 5 seconds have elapsed. If so, we break out the loop.

```
# Open a plain text file to store the retrieved sample data
text_file = open('sample_data.csv', 'w')

# Current tracker time
t_start = tk.trackerTime()

while True:
    # Break after 5 seconds have elapsed
    if tk.trackerTime() - t_start > 5000:
        break
```

Both sample and event data are timestamped according to the clock on the EyeLink Host PC. The time returned by *trackerTime*() is the same as that recorded in the EDF file. You can verify this by sending the time returned by *trackerTime*() as a message to the tracker.

Back to the script, a *while* loop is used to constantly check if a new sample is available. One way to tell if a sample is new is to compare its timestamp against the previous sample. If a sample is new, we save it to the text file.

```
    # Save gaze, HREF, raw, & pupil data to the plain text
    # file, if the sample is new
    if timestamp > smp_time:
        smp_data = map(str, [timestamp, gaze, href, raw, pupil])
        text_file.write('\t'.join(smp_data) + '\n')
        smp_time = timestamp
```

In this example, we log the samples into a plain text file, just for an illustration. It would not be sensible to log samples in this way in an actual experiment, as it creates a considerable overhead, and the data is already stored in the EDF file. Please see below for an excerpt from the output text file. The columns are sample timestamp, gaze position, HREF x/y, raw data, and pupil size (area or diameter in arbitrary units).

```
3577909.0 (872.0999755859375, 392.70001220703125) (1631.0, 7100.0)
(-1924.0, -3891.0) 366.0
3577910.0 (872.2000122070312, 393.3999938964844) (1632.0, 7093.0)
(-1924.0, -3889.0) 365.0
3577911.0 (872.2999877929688, 393.79998779296875) (1632.0, 7090.0)
(-1924.0, -3888.0) 365.0
3577912.0 (872.2000122070312, 397.20001220703125) (1631.0, 7060.0)
(-1924.0, -3876.0) 365.0
3577913.0 (872.0999755859375, 397.29998779296875) (1630.0, 7059.0)
(-1925.0, -3875.0) 365.0
3577914.0 (870.7000122070312, 397.3999938964844) (1619.0, 7059.0)
(-1930.0, -3875.0) 362.0
3577915.0 (870.7000122070312, 397.3999938964844) (1619.0, 7058.0)
(-1930.0, -3875.0) 362.0
```

An Example in PsychoPy: Gaze-Contingent Window

The following example script implements a simple gaze-contingent window task, taking advantage of the *Aperture* feature of PsychoPy. We use an *aperture* to reveal a rectangular region centered at the current gaze position (see Fig. 6.3).

Fig. 6.3 A simple gaze-contingent window task implemented in PsychoPy

```python
#!/usr/bin/env python3
#
# Filename: gaze_contingent_window.py
# Author: Zhiguo Wang
# Date: 2/6/2021
#
# Description:
# A gaze-contingent window task implemented in PsychoPy

import pylink
from psychopy import visual, core, event, monitors
from EyeLinkCoreGraphicsPsychoPy import EyeLinkCoreGraphicsPsychoPy

# Connect to the tracker
tk = pylink.EyeLink('100.1.1.1')

# Open an EDF data file
tk.openDataFile('psychopy.edf')

# Put the tracker in offline mode before we change tracking parameters
tk.setOfflineMode()

# Make all types of sample data available over the link
sample_flags = 'LEFT,RIGHT,GAZE,GAZERES,PUPIL,HREF,AREA,STATUS,INPUT'
tk.sendCommand(f'link_sample_data  = {sample_flags}')

# Screen resolution
SCN_W, SCN_H = (1280, 800)

# Open a PsyhocPy window with the "allowStencil" option
win = visual.Window((SCN_W, SCN_H), fullscr=False,
                    units='pix', allowStencil=True)

# Pass the display pixel coordinates (left, top, right, bottom) to the tracker
coords = f"screen_pixel_coords = 0 0 {SCN_W - 1} {SCN_H - 1}"
tk.sendCommand(coords)

# Request Pylink to use the custom EyeLinkCoreGraphicsPsychoPy library
# to draw calibration graphics (target, camera image, etc.)
genv = EyeLinkCoreGraphicsPsychoPy(tk, win)
pylink.openGraphicsEx(genv)

# Calibrate the tracker
calib_msg = visual.TextStim(win, text='Press ENTER to calibrate')
calib_msg.draw()
```

```
win.flip()
tk.doTrackerSetup()

# Set up an aperture and use it as a gaze-contingent window
gaze_window = visual.Aperture(win, shape='square', size=200)
gaze_window.enabled = True

# Load a background image to fill up the screen
img = visual.ImageStim(win, image='woods.jpg', size=(SCN_W, SCN_H))

# Put tracker in Offline mode before we start recording
tk.setOfflineMode()

# Start recording
tk.startRecording(1, 1, 1, 1)

# Cache some samples
pylink.msecDelay(100)

# show the image indefinitely until a key is pressed
gaze_pos = (-32768, -32768)
while not event.getKeys():
    # Check for new samples
    smp = tk.getNewestSample()
    if smp is not None:
        if smp.isRightSample():
            gaze_x, gaze_y = smp.getRightEye().getGaze()
        elif smp.isLeftSample():
            gaze_x, gaze_y = smp.getLeftEye().getGaze()

        # Draw the background image
        img.draw()

        # Update the window with the current gaze position
        gaze_window.pos = (gaze_x - SCN_W/2.0, SCN_H/2.0 - gaze_y)
        win.flip()

# Stop recording
tk.stopRecording()
# Put the tracker to offline mode
tk.setOfflineMode()
# Close the EDF data file on the Host
tk.closeDataFile()

# Download the EDF data file from Host
tk.receiveDataFile('psychopy.edf', 'psychopy.edf')
```

```
# Close the link to the tracker
tk.close()

# Close the graphics
win.close()
core.quit()
```

The code relevant to online sample retrieval is much like that in the previous example. The only thing new here is that we use the retrieved gaze data to update the aperture position. Note that, for the coordinates of the retrieved gaze data, the origin is the top-left corner of the screen. Some conversion is needed because the origin of the screen coordinates used in PsychoPy is the screen center.

```
# show the image indefinitely until a key is pressed
gaze_pos = (-32768, -32768)
while not event.getKeys():
    # Check for new samples
    smp = tk.getNewestSample()
    if smp is not None:
        if smp.isRightSample():
            gaze_x, gaze_y = smp.getRightEye().getGaze()
        elif smp.isLeftSample():
            gaze_x, gaze_y = smp.getLeftEye().getGaze()

        # Draw the background image
        img.draw()

        # Update the window with the current gaze position
        gaze_window.pos = (gaze_x - SCN_W/2.0, SCN_H/2.0 - gaze_y)
        win.flip()
```

Accessing Eye Events

As noted earlier, sample data is parsed online by the Host PC during recording to detect eye events, e.g., the onset and offset of saccades, fixations, and blinks. Depending on the tracker settings, both events and sample data can be saved in the EDF data files and made available over the link during testing. It is possible to convert the EDF data file into plain text files (see Chap. 8) to examine the various eye events parsed by the tracker. In the example data lines shown below, we see two pairs of events: SSACC (start of saccade) and ESACC (end of saccade) and SFIX (start of fixation) and EFIX (end of fixation). The letter "L" (or "R") following the event label indicates the data comes from the left or right eye. Note that for a "start" event, the timestamp is the only data recorded. For an "end" event, summary data is available, e.g., in the case of the EFIX event, the duration of fixation (2368 ms), the average x and y gaze position of the samples within the fixation (755, 447), and the average pupil size during the fixation (5161).

```
SSACC L 659818
ESACC L 659818 659962 148 749.0 413.0 752.0 437.0 0.35 571
SFIX L 659966
EFIX L 659966 662330 2368 755.0 447.0 5161
```

Eye events are useful in a number of research scenarios. For example, a saccade end event can be used to determine the response accuracy in a pro-/anti-saccade task and then conditionally administer subsequent trials. Online fixation events can be used to monitor whether participant fixed the proper region of the screen before showing the critical display or develop eye-typing applications that generate a key-press or mouse click events depending on where and how long participant looks on the screen.

Commands for Accessing Events

The Pylink library uses predefined constants to represent the types of eye events that can be retrieved online. For instance, if you enter "pylink.STARTBLINK" (without quotes) in the Python shell (after importing Pylink), the shell will return the corresponding constant "3." The Pylink constants for all eye events are listed in Table 6.1. Note that sample data is treated as a special type of "event"—SAMPLE_TYPE, represented by a constant (200).

```
>>> import pylink
>>> pylink.STARTBLINK
3
```

When talking about eye events, people are generally referring to fixations and saccades. However, as you can see from Table 6.1, these are not the only types of events you can retrieve during recording. For example, the Host PC also flags the start and end of blinks. The onset of blinks is flagged when the Host PC can no longer derive a valid X, Y location for gaze. Blink events are "wrapped" in (blink) saccades, i.e., the sequence of events is SSACC, SBLINK, EBLINK, ESACC. The reason for this is that as the eyelids descend over the pupil, from the camera's perspective, the pupil center is rapidly changing (shifting downward as the eyelid descends and obscures the upper part). The tracker will see a fast downward "movement" of the eyeball and report a saccade start event. When the eyelids reopen, the tracker sees the pupil center quickly stabilizes. The velocity of the perceived "movement" falls below the velocity threshold, leading to the detection of a saccade end event.

In the previous section, we illustrated how to read the latest sample as it arrives, with *getNewestSample()*. The EyeLink API also allows users to access data (sample and event) buffered in a queue in the same order they arrived. Buffering helps prevent data loss if the script fails to read the arrived data immediately; however, accessing data from a queue will introduce a delay if the queue contains significant amounts of data or cause data loss if the queue is not cleared quickly enough. To access event data in the queue, we first call *getNextData()* to check if there is an event there; this command will return a code representing an event (see Table 6.1). If this command returns 0, there is no data (sample or event) in the queue; otherwise, we call *getFloatData()* to grab the data out of the queue. The *getNextData*() and *getFloatData*() function pairs should be called as frequently as possible to empty the queue (e.g., within a *while* loop), so the event we retrieved is the latest one.

As noted, the access of eye events during testing is not real-time. When an eye event (e.g., STARTSACC and ENDFIX) is available over the link, it would have occurred some 20–40 ms ago. The short script below shows how to examine the event retrieval latency. We continuously examine the start and end events for fixations and saccades in a *while* loop. If an event is detected, we log a message that

Table 6.1 Pylink constants for various eye events

Variables	Constants	Description
STARTBLINK	3	Start of blink (pupil disappears)
ENDBLINK	4	End of blink (pupil reappears)
STARTSACC	5	Start of saccade
ENDSACC	6	End of saccade
STARTFIX	7	Start of fixation
ENDFIX	8	End of fixation
FIXUPDATE	9	Update fixation summary data during a fixation
MESSAGEEVENT	24	Message received by tracker
BUTTONEVENT	25	Button press
INPUTEVENT	28	Change of pin status on the Host parallel port
SAMPLE_TYPE	200	Sample data

contains the timestamp of the event (based on the Host PC clock). This message will have a timestamp. By comparing this message timestamp with the event timestamp, we can determine the delay between the event happening and it becoming available over the link.

```
#!/usr/bin/env python3
#
# Filename: event_retrieval.py
# Author: Zhiguo Wang
# Date: 5/26/2021
#
# Description:
# A short script illustrating online retrieval of eye events

import pylink


# Connect to the tracker
tk = pylink.EyeLink('100.1.1.1')

# Open an EDF data file on the Host PC
tk.openDataFile('ev_test.edf')

# Put the tracker in offline mode before we change tracking parameters
tk.setOfflineMode()

# Set sample rate to 1000 Hz
tk.sendCommand('sample_rate 1000')

# Make all types of event data are available over the link
event_flgs = 'LEFT,RIGHT,FIXATION,FIXUPDATE,SACCADE,BLINK,BUTTON,INPUT'
tk.sendCommand(f'link_event_filter = {event_flgs}')

# Open an SDL window for calibration
pylink.openGraphics()

# Set up the camera and calibrate the tracker
tk.doTrackerSetup()

# Put tracker in idle/offline mode before we start recording
tk.setOfflineMode()
```

```
# Start recording
tk.startRecording(1, 1, 1, 1)

# Wait for the block start event to arrive, give a warning
# if no event or sample is available
block_start = tk.waitForBlockStart(100, 1, 1)
if block_start == 0:
    print("ERROR: No link data received!")

# Check eye availability; 0-left, 1-right, 2-binocular
# read data from the right eye if tracking in binocular mode
eye_to_read = tk.eyeAvailable()
if eye_to_read == 2:
    eye_to_read = 1

# Get the current tracker time
t_start = tk.trackerTime()
while True:
    # Break after 5 seconds have elapsed
    if tk.trackerTime() - t_start > 5000:
        break

    # Retrieve the oldest event in the buffer
    dt = tk.getNextData()
    if dt in [pylink.STARTSACC, pylink.ENDSACC,
              pylink.STARTFIX, pylink.ENDFIX]:
        ev = tk.getFloatData()
        # Look for right eye events only; 0-left, 1-right
        if ev.getEye() == eye_to_read:
            # Send a message to the tracker when an event is
            # received over the link; include the timestamp
            # in the message to examine the link delay
            if dt == pylink.STARTSACC:
                tk.sendMessage(f'STARTSACC {ev.getTime()}')
            if dt == pylink.ENDSACC:
                tk.sendMessage(f'ENDSACC {ev.getTime()}')
            if dt == pylink.STARTFIX:
                tk.sendMessage(f'STARTFIX {ev.getTime()}')
            if dt == pylink.ENDFIX:
                tk.sendMessage(f'ENDFIX {ev.getTime()}')

# Stop recording
tk.stopRecording()

# Close the EDF data file on the Host
tk.closeDataFile()
```

```
# Download the EDF data file from Host
tk.receiveDataFile('ev_test.edf', 'ev_test.edf')

# Close the link to the tracker
tk.close()

# Close the window
pylink.closeGraphics()
```

As with sample data, we send over a command (*link_event_filter*) to control what types of event data can be accessed over the link during recording. More information about the event data flags (e.g., FIXATION, SACCADE, BLINK) that can be included in this command are available in the EyeLink user manuals.

```
# Make all types of event data are available over the link
event_flgs = 'LEFT,RIGHT,FIXATION,FIXUPDATE,SACCADE,BLINK,BUTTON,INPUT'
tk.sendCommand(f'link_event_filter = {event_flgs}')
```

A special "block start" event is passed over the link once the recording starts. This event contains information on what data will be available during recording, and we use *waitForBlockStart*() to look for this event in the link queue. The three arguments passed to this function specify how long to wait for the block start event and whether samples, events, or both types of data would be available during recording. This function returns 0 if no data is available over the link.

```
# Wait for the block start event to arrive, give a warning
# if no event or sample is available
block_start = tk.waitForBlockStart(100, 1, 1)
if block_start == 0:
    print("ERROR: No link data received!")
```

Once the block start event is received, we can use *eyeAvailable*() to check if data is available from the left eye, right eye, or both eyes. We need this piece of information to ensure the event data we retrieved is from the eye of interest to us (e.g., during binocular tracking).

```
# Check eye availability; 0-left, 1-right, 2-binocular
# read data from the right eye if tracking in binocular mode
eye_to_read = tk.eyeAvailable()
if eye_to_read == 2:
    eye_to_read = 1
```

At the end of the script, we download the EDF data file from the Host PC with *receiveDataFile*(). We then use the EDF converter provided by SR Research to convert the EDF data file into a plain text file (see Chapter 8 for details). We configure the EDF converter to extract only eye events and messages, ignoring the samples for simplicity.

In the file excerpt shown below, the first data line starts with the keyword "EFIX," marking the end of a fixation. This data line also contains some summary info about the same fixation, i.e., the start time, end time, duration, average gaze position, and average pupil size. This fixation ended at 5725698 ms, followed by a saccade onset event ("SSACC") at 5725699 ms. The third data line contains a message written by our script, running on the Display PC. It starts with the keyword "MSG," and its timestamp is 5725717 ms. This message was logged when our script received the EFIX event over the link. This EFIX event occurred at 5725698 ms; it is the same fixation summarized in the first data line. So, the link delay in retrieving the EFIX event is 5725717 – 5725698 = 19 ms. Using the script described above, the average link delay for saccade onset events (SSACC) is about 18 ms in my testing setup. The link delay for saccade offset events (ESACC) is about 35 ms, and that for fixation onset events (SFIX) is about 34 ms. These delays limit the utility of event data for gaze-contingent tasks. If you want gaze-contingents to happen as soon as possible, use sample data rather than event data.

```
EFIX R 5725568 5725698 131 560.5 294.0 335
SSACC R 5725699
MSG 5725717 ENDFIX 5725698.0
MSG 5725717 STARTSACC 5725699.0
ESACC R 5725699 5725753 55 556.1 291.1 122.6 129.5 13.10 451
SFIX R 5725754
MSG 5725788 ENDSACC 5725753.0
MSG 5725788 STARTFIX 5725754.0
```

You may want to retrieve more information from an eye event, in addition to its timestamp. The code snippet below shows how to retrieve the onset and offset time, amplitude, peak velocity, and start and end positions from an ESACC event.

```
# Current tracker time
t_start = tk.trackerTime()
while True:
    # Break after 5 seconds have elapsed
    if tk.trackerTime() - t_start > 5000:
        break

    # Retrieve the oldest event in the buffer
    dt = tk.getNextData()
    if dt == pylink.ENDSACC:
        ev = tk.getFloatData()
        # Look for right eye events only; 0-left, 1-right
        if ev.getEye() == eye_to_read:
            print('ENDSACC Event: \n',
                  'Amplitude', ev.getAmplitude(), '\n',
                  'Angle', ev.getAngle(), '\n',
                  'AverageVelocity', ev.getAverageVelocity(), '\n',
                  'PeakVelocity', ev.getPeakVelocity(), '\n',
                  'StartTime', ev.getStartTime(), '\n',
                  'StartGaze', ev.getStartGaze(), '\n',
                  'StartHREF', ev.getStartHREF(), '\n',
                  'StartPPD', ev.getStartPPD(), '\n',
                  'StartVelocity', ev.getStartVelocity(), '\n',
                  'EndTime', ev.getEndTime(), '\n',
                  'EndGaze', ev.getEndGaze(), '\n',
                  'EndHREF', ev.getEndHREF(), '\n',
                  'EndPPD', ev.getEndPPD(), '\n',
                  'EndVelocity', ev.getEndVelocity(), '\n',
                  'Eye', ev.getEye(), '\n',
                  'Time', ev.getTime(), '\n',
                  'Type', ev.getType(), '\n')
```

The output from the above code snippet is listed below. The saccade amplitude and the various velocity measures are reported in degrees of visual angles. The *getStartPPD*() and *getEndPPD*() functions return an estimate of how many pixels correspond to 1 degree of visual angle at the start and end of the saccade event. These measures all require a correct screen configuration, either stored on the Host PC or specified in the experimental script (see Chap. 4).

The Pylink commands that can be used to retrieve information from eye events are summarized in Table 6.2. For clarity, we group these commands based on their functionality:

```
ENDSACC Event:
 Amplitude (4.537407467464733, 0.5766124065721234)
 Angle -7.242339197477229
 AverageVelocity 102.0999984741211
 PeakVelocity 175.3000030517578
 StartTime 52627.0
 StartGaze (417.79998779296875, 1739.0999755859375)
 StartHREF (-1450.0, -2441.0)
 StartPPD (48.0, 49.70000076293945)
 StartVelocity 5.800000190734863
 EndTime 52671.0
 EndGaze (633.0999755859375, 1767.699951171875)
 EndHREF (-236.0, -2669.0)
 EndPPD (48.5, 49.20032176294065)
 EndVelocity 37.599998474121094
 Eye 1
 Time 52671.0
 Type 6
```

1. Commands available to all events. In this group of commands, *getTime()* returns the time point at which an event occurred (based on the Host PC clock), *getEye()* returns the eye being tracked (0-left, 1-right, 2-binocular), *getType()* returns the event type (i.e., the constants listed in Table 6.1), and *getStatus()* returns the errors or warnings related to an event if there were any.
2. These commands return the start and end time of an event.
3. These commands return eye position data in GAZE coordinates.
4. These commands return eye position data in HREF coordinates.
5. These commands return the resolution data (PPD, pixels per degree), i.e., how many screen pixels correspond to 1 degree of visual angle.
6. These commands return velocity data.
7. These commands return pupil size data.
8. Commands that are available to the ENDSACC event only; they return saccade amplitude and direction.

One thing worth mentioning here is that the fixation update (FIXUPDATE) event reports the same set of data as the fixation end event (ENDFIX), at predefined intervals (50 ms by default) following the onset of a fixation. In this way, it is possible to access the states of an (ongoing) fixation (e.g., the average gaze position) before it ends. This feature can be useful in HCI applications, e.g., using gaze to drive the mouse cursor.

Table 6.2 Pylink commands for retrieving event data. Here we use "X" to mark the properties available to each eye event

		STARTSACC	ENDSACC	STARTFIX	ENDFIX	STARTBLINK	ENDBLINK	FIXUPDATE
1	getTime()	X	X	X	X	X	X	X
	getEye()	X	X	X	X	X	X	X
	getType()	X	X	X	X	X	X	X
	getStatus()	X	X	X	X	X	X	X
2	getStartTime()	X	X	X	X	X	X	X
	getEndTime()		X		X		X	X
3	getStartGaze()	X	X	X	X			X
	getEndGaze()		X		X			X
	getAverageGaze()				X			X
4	getStartHREF()	X	X	X	X			X
	getEndHREF()		X		X			X
	getAverageHREF()				X			X
5	getStartPPD()	X	X	X	X			X
	getEndPPD()		X		X			X
6	getStartVelocity()	X	X	X	X			X
	getEndVelocity()		X		X			X
	getAverageVelocity()		X		X			X
	getPeakVelocity()		X		X			X
7	getStartPupilSize()			X	X			X
	getEndPupilSize()				X			X
	getAveragePupilSize()				X			X
8	getAmplitude()		X					
	getAngle()		X					

An Example in PsychoPy: Gaze Trigger

While the link delays limit the utility of event data for rapid gaze-contingent changes, the online parsed eye events can still be very useful in many research scenarios. For example, the ENDSACC event can be used to calculate saccadic response time in a visually guided saccade task within the task itself. In other words, the gaze data need not to be analyzed in a separate step; the script itself can calculate the key metric of saccade latency. The FIXUPDATE event, on the other hand, can be used to set up a "gaze trigger," so the task moves to the next step only if the participant has gazed at a given screen position for a certain amount of time (see the example script below).

```
#!/usr/bin/env python3
#
# Filename: gaze_trigger.py
# Author: Zhiguo Wang
# Date: 5/26/2021
#
# Description:
# A gaze trigger implemented in PsychoPy

import pylink
from psychopy import visual, core, event, monitors
from EyeLinkCoreGraphicsPsychoPy import EyeLinkCoreGraphicsPsychoPy
from math import hypot

# Connect to the tracker
tk = pylink.EyeLink('100.1.1.1')

# Open an EDF data file on the Host PC
tk.openDataFile('psychopy.edf')

# Put the tracker in offline mode before we change tracking parameters
tk.setOfflineMode()

# Make all types of eye events available over the link, especially the
# FIXUPDATE event, which reports the current status of a fixation at
# predefined intervals (default = 50 ms)
event_flags = 'LEFT,RIGHT,FIXATION,FIXUPDATE,SACCADE,BLINK,BUTTON,INPUT'
tk.sendCommand(f'link_event_filter = {event_flags}')

# Screen resolution
SCN_W, SCN_H = (1280, 800)

# Open a PsyhocPy window
win = visual.Window((SCN_W, SCN_H), fullscr=False, units='pix')
```

```
# Pass the display pixel coordinates (left, top, right, bottom) to the tracker
coords = f"screen_pixel_coords = 0 0 {SCN_W - 1} {SCN_H - 1}"
tk.sendCommand(coords)

# Request Pylink to use the custom EyeLinkCoreGraphicsPsychoPy library
# to draw calibration graphics (target, camera image, etc.)
genv = EyeLinkCoreGraphicsPsychoPy(tk, win)
pylink.openGraphicsEx(genv)

# Calibrate the tracker
calib_msg = visual.TextStim(win, text='Press ENTER twice to calibrate')
calib_msg.draw()
win.flip()
tk.doTrackerSetup()

# Run 3 trials in a for-loop
# in each trial, first show a fixation dot, wait for the participant
# to gaze at the fixation dot, then present an image for 2 secs
for i in range(3):
    # Prepare the fixation dot in memory
    fix = visual.GratingStim(win, tex='None', mask='circle', size=30.0)

    # Load the image
    img = visual.ImageStim(win, image='woods.jpg', size=(SCN_W, SCN_H))

    # Put tracker in Offline mode before we start recording
    tk.setOfflineMode()

    # Start recording
    tk.startRecording(1, 1, 1, 1)

    # Wait for the block start event to arrive, give a warning
    # if no event or sample is available
    block_start = tk.waitForBlockStart(100, 1, 1)
    if block_start == 0:
        print("ERROR: No link data received!")

    # Check eye availability; 0-left, 1-right, 2-binocular
    # read data from the right eye if tracking in binocular mode
    eye_to_read = tk.eyeAvailable()
    if eye_to_read == 2:
        eye_to_read = 1

    # Show the fixation dot
    fix.draw()
    win.flip()
```

```
    # Gaze trigger
    # wait for gaze on the fixation dot (for a minimum of 300 ms)
    fix_dot_x, fix_dot_y = (SCN_W/2.0, SCN_H/2.0)
    triggered = False
    fixation_start_time = -32768
    while not triggered:
        # Check if any new events are available
        dt = tk.getNextData()
        if dt == pylink.FIXUPDATE:
            ev = tk.getFloatData()
            if ev.getEye() == eye_to_read:
                # 1 deg = ? pixels in the current fixation
                ppd_x, ppd_y = ev.getEndPPD()

                # Get the gaze error
                gaze_x, gaze_y = ev.getAverageGaze()
                gaze_error = hypot((gaze_x - fix_dot_x)/ppd_x,
                                   (gaze_y - fix_dot_y)/ppd_y)

                if gaze_error < 1.5:
                    # Update fixation_start_time, following the first
                    # FIXUPDATE event
                    if fixation_start_time < 0:
                        fixation_start_time = ev.getStartTime()
                    else:
                        # Break if the gaze is on the fixation dot
                        # for > 300 ms
                        if (ev.getEndTime() - fixation_start_time) >= 300:
                            triggered = True
                else:
                    fixation_start_time = -32768

    # Show the image for 2 secs
    img.draw()
    win.flip()
    core.wait(2.0)

    # Clear the screen
    win.color = (0, 0, 0)
    win.flip()
    core.wait(0.5)

    # Stop recording
    tk.stopRecording()

# Close the EDF data file on the Host
tk.closeDataFile()
```

```
# Download the EDF data file from Host
tk.receiveDataFile('psychopy.edf', 'psychopy.edf')

# Close the link to the tracker
tk.close()

# Close the graphics
win.close()
core.quit()
```

This script is similar to the gaze-contingent window task presented in the previous section. One thing worth mentioning here is the use of the FIXUPDATE event. This special eye event allows users to access information about the currently ongoing fixation before it ends. Following the onset of a fixation, the tracker reports information about the ongoing fixation at predefined intervals (50 ms by default) by making a new FIXUPDATE event available over the link. The FIXUPDATE event returns the same set of data as the FIXEND event. In the example script, we use the *getAverageGaze*() command to retrieve the average gaze position during the fixation update interval. We then check if the average gaze position is close to the fixation dot, taking advantage of the resolution data that can be accessed via the FIXUPDATE event. We break out the while loop if gaze is close to the fixation dot (< 1.5 degrees) for more than 300 ms.

```
    # 1 deg = ? pixels in the current fixation
    ppd_x, ppd_y = ev.getEndPPD()

    # Get the gaze error
    gaze_x, gaze_y = ev.getAverageGaze()
    gaze_error = hypot((gaze_x - fix_dot_x)/ppd_x,
                       (gaze_y - fix_dot_y)/ppd_y)
    if gaze_error < 1.5:
        # Update fixation_start_time, following the first
        # FIXUPDATE event
        if fixation_start_time < 0:
            fixation_start_time = ev.getStartTime()
        else:
            # Break if the gaze is on the fixation dot
            # for > 300 ms
            if (ev.getEndTime() - fixation_start_time) >= 300:
                triggered = True
    else:
        fixation_start_time = -32768
```

To briefly recap, the EyeLink tracker images the eye(s) up to 2000 times per second, and each eye image is analyzed to estimate the eye (or gaze) position. In addition to the raw eye position data (samples), the tracker can also detect and record eye events during recording. This chapter illustrates how to access the sample and event data over the Ethernet link during recording. In the following chapters, we will cover some more advanced topics on the Pylink library (Chap. 7) and discuss data analysis and visualization in Python (Chap. 8).

Chapter 7
Advanced Use of Pylink

Contents

Chapters 4, 5, and 6 discussed the frequently used features of Pylink, the Python wrapper for the EyeLink API. In this chapter, we will discuss three advanced topics: (a) how to send drawing commands or images to the Host PC to show a backdrop during recording, (b) how to use the Host PC parallel port to relay TTLs to third-party devices, and (c) how to customize the graphics routines used during tracker calibration. The first topic is useful in many research scenarios; however, the second and third topics are safe to skip unless you are developing an advanced eye-tracking application.

Drawing to the Host PC

It can be helpful to have some background graphics on the Host PC screen during data recording. We can then examine gaze data in real-time to detect tracking problems (e.g., the gaze is always above or below the text lines in a reading task) and monitor the subject task performance. The Pylink library offers three possible solutions, which we will discuss in turn.

Draw Simple Graphics on the Host

The first solution is to send "draw" commands to the Host PC. These draw commands will draw simple graphics, like a rectangle or a fixation cross, on the Host PC screen in recording mode. The draw commands supported by the Host PC are documented in

Z. Wang, *Eye-Tracking with Python and Pylink*,
https://doi.org/10.1007/978-3-030-82635-2_7

the COMMANDS.INI file in the /elcl/exe folder on the Host PC. For instance, the *clear_screen* command will clear the Host screen, and the *draw_line* command will draw a line. The Host PC supports a limited number of colors. The colors available on the Host PC and their corresponding color codes are listed in Table 7.1. The following draw command will clear the Host PC screen and fill it with black (color code 0).

```
tk = pylink.EyeLink()
tk.sendCommand('clear_screen 0')
```

The draw commands supported by the Host PC are also available as Pylink functions. So, calling the Pylink function *clearScreen(0)* will also clear the Host display and fill it with black. See Table 7.2 for a list of the Host draw commands and their corresponding Pylink wrapper functions.

```
tk.clearScreen(0)
```

The draw commands are useful when you would like to mark a small number of essential locations on the screen, e. g., the correct and incorrect saccade target locations in an anti-saccade task or the trajectory and limits of a smooth pursuit target.

The bitmapBackdrop() Function

The *bitmapBackdrop()* function sends an image to the Host PC to use as the backdrop. This command is supported by all versions of the EyeLink Host PC. However, this function can be slow, and it is not recommended for timing-critical tasks (e.g.,

Table 7.1 The colors supported by the Host PC and the corresponding codes

0—black	4—red	8—dark gray	12—light red
1—blue	5—magenta	9—light blue	13—bright magenta
2—green	6—brown	10—light green	14—yellow
3 —cyan	7—light gray	11—light cyan	15—bright white

Table 7.2 Host commands for simple drawing and their corresponding Pylink wrapper functions

Pylink wrapper functions	Commands supported by the Host
echo()	echo <text>
drawText()	draw_text <x1> <y1> <color> <text>
drawLine()	draw_line <x1> <y1> <x2> <y2> <color>
drawFilledBox()	draw_filled_box <x1> <y1> <x2> <y2> <color>
drawCross()	draw_cross <x> <y> <color>
drawBox()	draw_box <x1> <y1> <x2> <y2> <color>
clearScreen()	clear_screen <color>

an event-related fMRI task). When using this function, it should be called at the beginning of a trial, before data recording starts. The parameters for this function are discussed below for completeness. Novice users can skip the details and use the example code provided at the end of this section.

```
bitmapBackdrop(<iwidth>, <iheight>, <pixels>, <crop_x>, <crop_y>, <crop_
width>, <crop_height>, <host_x>, <host_y>, <xfer_option>)
```

The <iwidth> and <iheight> parameters specify the image width and height. The <pixels> parameter is the original image pixels formatted as a two-dimensional array (will explain later). The <crop_*> parameters are useful if you need to show a portion of the image on the Host screen rather than the entire image. The <host_x> and <host_y> parameters control the position of the image on the Host. The <xfer_options> parameter determines the appearance of the image on the Host; the default is BX_MAXCONTRAST, i.e., it maximizes the contrast to get the clearest image.

The <pixel> parameter requires a two-dimensional array containing the original image pixels. The pixels of each image line are put in a list, e.g., [pix_1, pix_2, ..., pix_n]; the lines that constitute an image are put into another list, e.g., [line_1, line_2, …, line_n]. Each image pixel is represented as a tuple of RGB values, i.e., (R, G, B), or a hexadecimal value specifying the alpha channel and the RGB values, i.e., 0xAARRGGBB. For instance, you may construct a checkerboard with 6 x 6 pixels (see Fig. 7.1) with a <pixel> list of RGB values in the following format.

```
# Construct an image in <pixel> format with RGB tuples
white_rgb = (255, 255, 255)
black_rgb = (0, 0, 0)
pixels_rgb = [
    [black_rgb, black_rgb, black_rgb, white_rgb, white_rgb, white_rgb],
    [black_rgb, black_rgb, black_rgb, white_rgb, white_rgb, white_rgb],
    [black_rgb, black_rgb, black_rgb, white_rgb, white_rgb, white_rgb],
    [white_rgb, white_rgb, white_rgb, black_rgb, black_rgb, black_rgb],
    [white_rgb, white_rgb, white_rgb, black_rgb, black_rgb, black_rgb],
    [white_rgb, white_rgb, white_rgb, black_rgb, black_rgb, black_rgb],
    ]
```

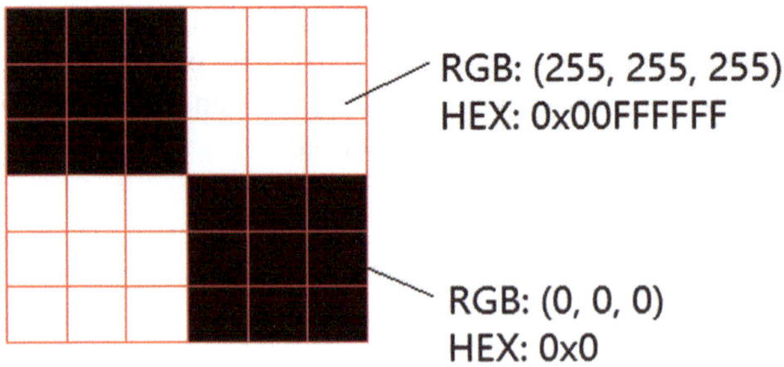

Fig. 7.1 A 6 × 6-pixel image. The <pixel> format supported by the bitmapBackdrop() function requires a two-dimensional list that contains all the pixels from this image

The same <pixel> list with HEX values will give you the same checkerboard.

```
# construct an image in <pixel> format with hexadecimal values alpha-R-G-B
white_hex = 0x00FFFFFF
black_hex = 0x0
pixels_hex = [
    [black_hex, black_hex, black_hex, white_hex, white_hex, white_hex],
    [black_hex, black_hex, black_hex, white_hex, white_hex, white_hex],
    [black_hex, black_hex, black_hex, white_hex, white_hex, white_hex],
    [white_hex, white_hex, white_hex, black_hex, black_hex, black_hex],
    [white_hex, white_hex, white_hex, black_hex, black_hex, black_hex],
    [white_hex, white_hex, white_hex, black_hex, black_hex, black_hex],
    ]
```

There is no Pylink function for converting an image into the <pixel> format supported by the *bitmapBackdrop*() function. Nevertheless, with the Python Image Library (PIL), this conversion requires only a few lines of code.

In the example code below, we open the image with the Image module from the Python Image Library (PIL), get the width and height, and then create a pixel access object named "pixels" by calling the *load()* method. We then reformat the image pixels into a two-dimensional array with the list comprehension trick (see Chap. 1). Then, we call the *bitmapBackdrop()* command to show the image on the Host PC screen (Fig. 7.2).

```
#!/usr/bin/env python3
#
# Filename: bitmap_backdrop.py
# Author: Zhiguo Wang
# Date: 4/27/2021
#
# Description:
# Transfer an image to the Host to use as the backdrop

import pylink
from PIL import Image

# Connect to the tracker
tk = pylink.EyeLink('100.1.1.1')

# Pass display dimension (left, top, right, bottom) to the tracker
tk.sendCommand('screen_pixel_coords = 0 0 1023 767')

# Put the tracker in offline mode before we transfer the image
tk.setOfflineMode()

# convert the image to the <pixel> format supported by
# the bitmapBackdrop() command
im = Image.open('quebec.jpeg')  # open an image with PIL
w, h = im.size  # get the width and height of the image
pixels = im.load()  # access the pixel data
# reformat the pixels
pixels_img = [[pixels[i, j] for i in range(w)] for j in range(h)]

# Transfer the images to the Host PC screen
tk.sendCommand('clear_screen 0')
tk.sendCommand('echo PIXELs_FROM_IMAGE')
tk.bitmapBackdrop(w, h, pixels_img, 0, 0, w, h,
                  50, 50, pylink.BX_MAXCONTRAST)

# Show the image for 3-sec on the Host PC
pylink.msecDelay(3000)

# Clear up the Host screen
tk.sendCommand('clear_screen 0')

# Close the connection
tk.close()
```

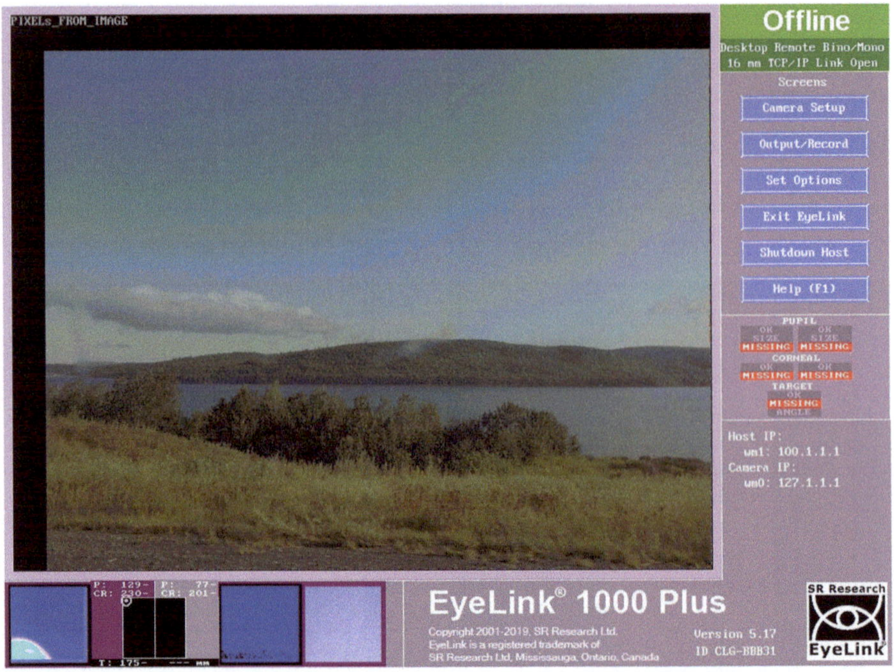

Fig. 7.2 Draw an image on the Host PC and shift it to (50, 50) with the bitmapBackdrop() function

Converting an image to the <pixel> format with the list comprehension trick above takes about 30 ms on a decent machine. In my setup, the execution of the *bitmapBackdrop()* function itself takes ~300–400 ms for an image that is 1920 x 1080 pixels and about 85 ms for an image that is 800 x 600 pixels. As noted, this timing overhead may be problematic for some experimental tasks, and you should always call the *bitmapBackdrop()* function before data recording starts.

The imageBackdrop() Function

The latest version of the EyeLink Developer's Kit has introduced a new function for sending an image to the Host PC to use as a backdrop, i.e., *imageBackrop*().

```
imageBackdrop(<image_file>, <crop_x>, <crop_y>, <crop_width>, <crop_
height>, <host_x>, <host_y>, <xfer_option>)
```

One advantage of the *imageBackdrop*() command is that it does not require any image format conversion. The first parameter of this command is the image file. The rest of the parameters are the same as the *bitmapBackdrop*() command, except that

```
tk.imageBackdrop('quebec.jpeg', 0, 0, 1024, 768, 0, 0, pylink.BX_
MAXCONTRAST)
```

the width and height of the original image are no longer required. One downside of this command is that it does not scale the image on the Host PC.

Sending TTLs via the Host PC

TTL stands for transistor-transistor logic, a logic unit with two transistors to ensure the output is either low (<0.8 V) or high (>2.4 V). The output (high and low) is converted into digital signals 1 and 0, enabling the transition of data between systems (e.g., a PC and a printer). Nowadays, the term TTL often refers to the signal itself. Computers send TTLs via the parallel port, which usually comes with 25 pins (sometimes referred to as D-Sub25 or DB25). There are four types of pins: Data, Status, Control, and Ground (see Fig. 7.3). The Data, Status, and Control pins are known as registers, which have addresses in computer memory (the Data Register is usually 0x378, the Status Register is usually 0x379, and the Control register is usually 0x37A). By writing a "byte" (e.g., 10000000), we can change the output on the eight Data pins to low or high to send a value between 0 (0x0) and 255 (0xFF) to another device.

The co-registration of eye movement and EEG recordings allows researchers to perform ICA-based artifact rejection and event-related analysis time-locked to eye events, e.g., fixation-related potentials (FRPs). There are multiple ways to ensure that the EEG and eye movement data contain common "sync signals," so that they

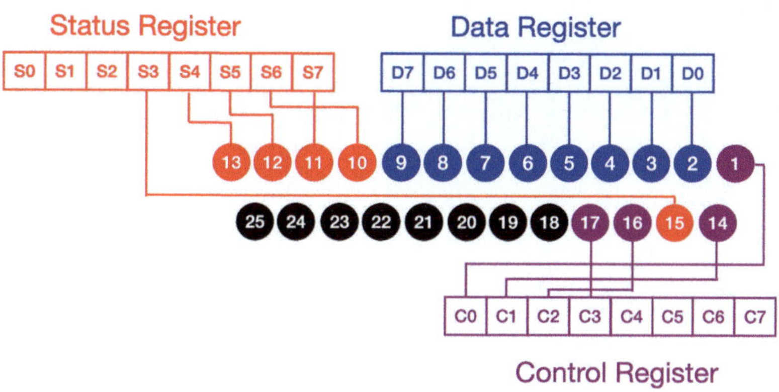

Fig. 7.3 The pin layout of a DB25 parallel port

can be aligned in time during offline analysis. If your stimulus presentation PC has a parallel port, you may use a Y-shape splitter cable to send the same TTL signal to both the EEG recording device and the EyeLink Host PC. If there is no parallel port on the stimulus presentation PC, you can send over a command to request the EyeLink Host PC to record and relay TTL signals to the EEG device. The EyeLink Host command for this task is "write_ioport," which is wrapped in the Pylink *writeIOPort()* function. The *writeIOPort()* function takes two arguments:

- < ioport>—address of the parallel port to write to. The base address for the onboard parallel port card on the EyeLink II/1000 Host PC is 0x378, and that on the EyeLink 1000 Plus Host PC is 0x8.
- <data>—the byte to write, e.g., 15 in decimal or 0xF in hexadecimal format.

```
#!/usr/bin/env python3
#
# Filename: TTL_through_host.py
# Author: Zhiguo Wang
# Date: 4/27/2021
#
# Description:
# Sending TTLs through the EyeLink Host PC
# Here we assume an EyeLink 1000 Plus tracker is being tested, so the
# base address is 0x8. For EyeLink 1000, the base address is 0x378

import pylink

# Connect to the tracker
tk = pylink.EyeLink('100.1.1.1')

# Open and EDF data file on the Host
tk.openDataFile('ttl_test.edf')

# Start recording
tk.startRecording(1, 1, 1, 1)

# Using the Pylink function writeIOPort to send TTLs

# Clear the Data Register
tk.writeIOPort(0x8, 0)
pylink.pumpDelay(100)

for i in range(201, 209):
    # Write a TTL to the Data Register
    tk.writeIOPort(0x8, i)
```

```
    # TTL signal duration--20 ms
    pylink.pumpDelay(20)
    # Clear the Data Register
    tk.writeIOPort(0x8, 0)
    # Wait for 1 second before we send the next TTL
    pylink.pumpDelay(1000)

# Using the Host 'write_ioport' command to send TTLs
# The "*" in the command request the Host to log the command in
# the EDF data file

# Clear the Data Register
tk.sendCommand('write_ioport 0x8 0')
pylink.pumpDelay(100)

for j in range(1, 9):
    # Write a TTL to the Data Register
    tk.sendCommand(f'*write_ioport 0x8 {j}')
    # TTL signal duration--20 ms
    pylink.pumpDelay(20)
    # Clear the Data Register
    tk.sendCommand('write_ioport 0x8 0')
    # Wait for 1 second before we send the next TTL
    pylink.pumpDelay(1000)

# Stop recording
tk.stopRecording()

# Close the EDF data file and download it from the Host PC
tk.closeDataFile()
tk.receiveDataFile('ttl_test.edf', 'ttl_test.edf')

# Close the link to the tracker
tk.close()
```

The above script shows how to send TTLs through the Host PC with either the Pylink *writeIOPort*() function or the Host command '*write_ioport*'. The latter solution is preferred as it allows us to log the command in the EDF data file.

Before we send any TTLs, we first clear the Data Register. Note that I added an asterisk to the '*write_ioport*' Host command. This additional flag will request the

EyeLink Host to log this command and relay the TTL to other devices. Example data lines recorded by the EyeLink Host PC are shown in the box below.

```
MSG    264631 !CMD 0 write_ioport 0x8 0
MSG    264641 !CMD 0 write_ioport 0x8 7
MSG    265641 !CMD 0 write_ioport 0x8 0
```

The eye movement and EEG co-registration methods discussed so far are "TTL-only" solutions. You can also send messages over the Ethernet link to the Host PC and then send the matching TTLs to the EEG device. In offline analysis, use the matching TTLs and messages to co-register the data. It is beyond the scope of this book to discuss the various co-registration solutions. An excellent tutorial is available from the EYE-EEG toolbox website (http://www2.hu-berlin.de/eyetracking-eeg/tutorial.html).

Calibration and Custom CoreGraphics

This section will present some technical details about the EyeLink calibration routine. As an advanced topic, it is safe for novice users to skip it. Still, it can be useful to those who would like to have more control over the calibration process, e.g., replacing the default calibration targets with a picture.

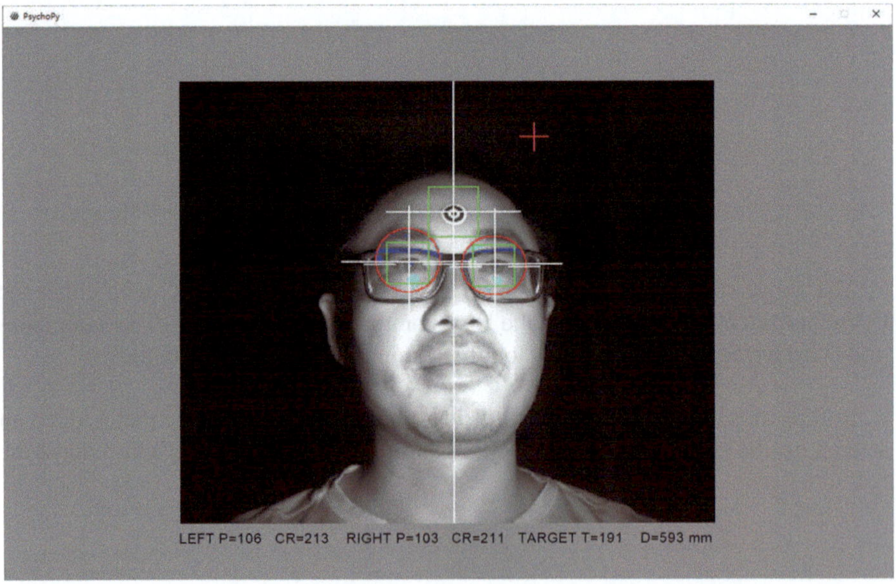

Fig. 7.4 A custom CoreGraphics library also handles the drawing of the camera image

The EyeLink Host PC controls the calibration routine. When running an experimental script on the Display PC, a command is sent to the Host PC to trigger the calibration routine. The drawing of the calibration target on the Display PC is handled by the EyeLink API, which listens to the Host PC for instructions on where to draw the next calibration target. The EyeLink API comes with graphics routines for drawing the calibration target on the Display PC, etc. These graphics routines are built on the SDL library, and we call the *pylink.openGraphics()* command to evoke them. The EyeLink API also allows users to customize these graphics routines with a library other than SDL. For convenience, we will refer to these graphics routines as "CoreGraphics." In this section, I will discuss the various functions of a custom CoreGraphics library implemented in PsychoPy.

Before diving in, note that a CoreGraphics library performs more tasks than drawing the calibration target. It also handles the drawing of the camera image (see Fig. 7.4) and the registration of keyboard and mouse input on the Display PC side. With the camera image available on the Display PC screen, the experimenter can easily tell if pupil and CR thresholds are reasonable and if the camera is focused without looking at the Host PC screen. This feature is especially helpful if the Host PC is in a separate control room.

The details of the various custom CoreGraphics routines are not important to ordinary users. You do not need to call these routines directly; you only need to request Pylink to use a custom CoreGraphics library. To use the default CoreGraphics library provided in Pylink, we call *pylink.openGraphics()*. To use a custom CoreGraphics library, we call *pylink.openGraphicsEx()* instead. Here, "Ex" stands for "external," as opposed to "built-in."

```
#!/usr/bin/env python3
#
# Filename: demo.py
# Author: Zhiguo Wang
# Date: 2/4/2021
#
# Description:
# This short script shows how to request Pylink to use the
# EyeLinkCoreGraphcicPsychoPy library

import pylink
from psychopy import visual, core, event, monitors
from EyeLinkCoreGraphicsPsychoPy import EyeLinkCoreGraphicsPsychoPy

# Connect to the tracker
tk = pylink.EyeLink('100.1.1.1')

# Open a PsychoPy window
SCN_WIDTH, SCN_HEIGHT = (1280, 800)
```

```
win = visual.Window((SCN_WIDTH, SCN_HEIGHT),
                    fullscr=False,
                    units='pix')

# Pass display dimension (left, top, right, bottom) to the tracker
coords = f"screen_pixel_coords = 0 0 {SCN_WIDTH - 1} {SCN_
HEIGHT - 1}"
tk.sendCommand(coords)

# Create a custom graphics environment (genv) for calibration
genv = EyeLinkCoreGraphicsPsychoPy(tk, win)
pylink.openGraphicsEx(genv)

# Calibrate the tracker
# when a gray screen comes up, press Enter to show the camera image
# press C to calibrate, V to validate, O to exit the calibration routine
tk.doTrackerSetup()

# Close the connection to the tracker
tk.close()

# Quit PsychoPy
win.close()
core.quit()
```

In the example script above, we first import the CoreGraphics library (module) implemented in PsychoPy, i.e., *EyeLinkCoreGraphicsPsychoPy*. We will give a brief discussion on "class" later in this chapter. For now, it is sufficient to know that *EyeLinkCoreGraphicsPsychoPy* is a Python class that helps to configure and execute the calibration routines.

```
from EyeLinkCoreGraphicsPsychoPy import EyeLinkCoreGraphicsPsychoPy
```

We connect to the tracker (*tk*) and open a PsychoPy window (*win*). Then, we call *EyeLinkCoreGraphicsPsychoPy()* to set up a custom graphics environment (*genv*) by passing two arguments to it, i.e., the tracker connection (*tk*) and the PsychoPy window (*win*).

```
# Create a custom graphics environment (genv) for calibration
genv = EyeLinkCoreGraphicsPsychoPy(tk, win)
```

We then call *pylink.openGraphicsEx()* to request Pylink to use the custom graphics environment (*genv*).

```
pylink.openGraphicsEx(genv)
```

A basic but functioning custom CoreGraphics library implemented in PsychoPy is presented below. A more feature-rich implementation is included in the EyeLink Developer's Kit. Here, I will briefly discuss the purpose of the various functions we need to define in a custom CoreGraphics library.

```
#!/usr/bin/env python3
#
# Filename: EyeLinkCoreGraphicsPsychoPy.py
# Author: Zhiguo Wang
# Date: 2/4/2021
#
# Description:
# An EyeLink coregraphics library (calibration routine)
# for PsychoPy experiments.

import os
import platform
import array
import string
import pylink
from psychopy import visual, event, core
from math import sin, cos, pi
from PIL import Image, ImageDraw
from psychopy.sound import Sound

class EyeLinkCoreGraphicsPsychoPy(pylink.EyeLinkCustomDisplay):
    def __init__(self, tracker, win):
        '''Initialize
```

```
        tracker: an EyeLink instance (connection)
        win: the PsychoPy window we use for calibration'''

        pylink.EyeLinkCustomDisplay.__init__(self)

        # background and target color
        self._backgroundColor = win.color
        self._foregroundColor = 'black'

        # window to use for calibration
        self._display = win
        # make the mouse cursor invisible
        self._display.mouseVisible = False

        # display width & height
        self._w, self._h = win.size

        # resolution fix for Mac retina displays
        if 'Darwin' in platform.system():
            sys_cmd = 'system_profiler SPDisplaysDataType | grep Retina'
            is_ret = os.system(sys_cmd)
            if is_ret == 0:
                self._w = int(self._w / 2.0)
                self._h = int(self._h / 2.0)

        # store camera image pixels in an array
        self._imagebuffer = array.array('I')

        # store the color palette for camera image drawing
        self._pal = None

        # initial size of the camera image
        self._size = (384, 320)

        # initial mouse configuration
        self._mouse = event.Mouse(False)
        self.last_mouse_state = -1

        # camera image title
        self._msgHeight = self._size[1]/16.0
        self._title = visual.TextStim(self._display, '',
                                      wrapWidth=self._w,
                                      color=self._foregroundColor)
```

```
        # calibration target
        self._targetSize = self._w/64.
        self._tar = visual.Circle(self._display,
                                  size=self._targetSize,
                                  lineColor=self._foregroundColor,
                                  lineWidth=self._targetSize/2)

        # calibration sounds (beeps)
        self._target_beep = Sound('type.wav', stereo=True)
        self._error_beep = Sound('error.wav', stereo=True)
        self._done_beep = Sound('qbeep.wav', stereo=True)

        # a reference to the tracker connection
        self._tracker = tracker

        # for a clearer view we always enlarge the camera image
        self.imgResize = None

    def setup_cal_display(self):
        '''Set up the calibration display '''

        self._display.clearBuffer()


    def clear_cal_display(self):
        '''Clear the calibration display'''

        self._display.color = self._backgroundColor
        self._display.flip()

    def exit_cal_display(self):
        '''Exit the calibration/validation routine'''

        self.clear_cal_display()


    def record_abort_hide(self):
        '''This function is called if aborted'''

        pass


    def erase_cal_target(self):
        '''Erase the target'''

        self.clear_cal_display()
```

```
    def draw_cal_target(self, x, y):
        '''Draw the target'''

        self.clear_cal_display()

        # target position
        xVis = (x - self._w/2.0)
        yVis = (self._h/2.0 - y)

        # draw the calibration target
        self._tar.pos = (xVis, yVis)
        self._tar.draw()
        self._display.flip()

    def play_beep(self, beepid):
        ''' Play a sound during calibration/drift-correction.'''

        if beepid in [pylink.CAL_TARG_BEEP, pylink.DC_TARG_BEEP]:
            self._target_beep.play()
        elif beepid in [pylink.CAL_ERR_BEEP, pylink.DC_ERR_BEEP]:
            self._error_beep.play()
        elif beepid in [pylink.CAL_GOOD_BEEP, pylink.DC_GOOD_BEEP]:
            self._done_beep.play()
        core.wait(0.4)

    def getColorFromIndex(self, colorindex):
       '''Retrieve the colors for camera image elements, e.g., crosshair'''

        if colorindex == pylink.CR_HAIR_COLOR:
            return (255, 255, 255)
        elif colorindex == pylink.PUPIL_HAIR_COLOR:
            return (255, 255, 255)
        elif colorindex == pylink.PUPIL_BOX_COLOR:
            return (0, 255, 0)
        elif colorindex == pylink.SEARCH_LIMIT_BOX_COLOR:
            return (255, 0, 0)
        elif colorindex == pylink.MOUSE_CURSOR_COLOR:
            return (255, 0, 0)
        else:
            return (128, 128, 128)

    def draw_line(self, x1, y1, x2, y2, colorindex):
        '''Draw a line '''

        color = self.getColorFromIndex(colorindex)

        # scale the coordinates
        w, h = self._img.im.size
        x1 = int(x1 / 192 * w)
```

```
        x2 = int(x2 / 192 * w)
        y1 = int(y1 / 160 * h)
        y2 = int(y2 / 160 * h)

        # draw the line
        if not any([x < 0 for x in [x1, x2, y1, y2]]):
            self._img.line([(x1, y1), (x2, y2)], color)

    def draw_lozenge(self, x, y, width, height, colorindex):
        ''' draw a lozenge to show the defined search limits '''

        color = self.getColorFromIndex(colorindex)

        # scale the coordinates
        w, h = self._img.im.size
        x = int(x / 192 * w)
        y = int(y / 160 * h)
        width = int(width / 192 * w)
        height = int(height / 160 * h)

        # draw the lozenge
        if width > height:
            rad = int(height / 2.)
            if rad == 0:
                return
            else:
                self._img.line([(x + rad, y), (x + width - rad, y)], color)
                self._img.line([(x + rad, y + height),
                                (x + width - rad, y + height)], color)
                self._img.arc([x, y, x + rad*2, y + rad*2], 90, 270, color)
                self._img.arc([x + width - rad*2, y, x + width, y + height],
                               270, 90, color)
        else:
            rad = int(width / 2.)
            if rad == 0:
                return
            else:
                self._img.line([(x, y + rad), (x, y + height - rad)], color)
                self._img.line([(x + width, y + rad),
                                (x + width, y + height - rad)], color)
                self._img.arc([x, y, x + rad*2, y + rad*2], 180, 360, color)
                self._img.arc([x, y + height-rad*2, x + rad*2, y + height],
                               0, 180, color)

    def get_mouse_state(self):
        '''Get the current mouse position and status'''

        w, h = self._display.size
        X, Y = self._mouse.getPos()
```

```
        # scale the mouse position, so the cursor stays on the camera image
        mX = (X + w/2.0)/w*self._size[0]/2.0
        mY = (h/2.0 - Y)/h*self._size[1]/2.0

        state = self._mouse.getPressed()[0]

        return ((mX, mY), state)


    def get_input_key(self):
        '''This function is repeatedly pooled to check
        keyboard events'''

        ky = []
        for keycode, modifier in event.getKeys(modifiers=True):
            k = pylink.JUNK_KEY
            if keycode == 'f1': k = pylink.F1_KEY
            elif keycode == 'f2': k = pylink.F2_KEY
            elif keycode == 'f3': k = pylink.F3_KEY
            elif keycode == 'f4': k = pylink.F4_KEY
            elif keycode == 'f5': k = pylink.F5_KEY
            elif keycode == 'f6': k = pylink.F6_KEY
            elif keycode == 'f7': k = pylink.F7_KEY
            elif keycode == 'f8': k = pylink.F8_KEY
            elif keycode == 'f9': k = pylink.F9_KEY
            elif keycode == 'f10': k = pylink.F10_KEY
            elif keycode == 'pageup': k = pylink.PAGE_UP
            elif keycode == 'pagedown': k = pylink.PAGE_DOWN
            elif keycode == 'up': k = pylink.CURS_UP
            elif keycode == 'down': k = pylink.CURS_DOWN
            elif keycode == 'left': k = pylink.CURS_LEFT
            elif keycode == 'right': k = pylink.CURS_RIGHT
            elif keycode == 'backspace': k = ord('\b')
            elif keycode == 'return': k = pylink.ENTER_KEY
            elif keycode == 'space': k = ord(' ')
            elif keycode == 'escape': k = 27
            elif keycode == 'tab': k = ord('\t')
            elif keycode in string.ascii_letters:
                k = ord(keycode)
            elif k == pylink.JUNK_KEY:
                k = 0

            # plus & minus signs for CR adjustment
            if keycode in ['num_add', 'equal']:
                k = ord('+')
            if keycode in ['num_subtract', 'minus']:
                k = ord('-')
```

```
            # handles key modifier
            if modifier['alt'] is True: mod = 256
            elif modifier['ctrl'] is True: mod = 64
            elif modifier['shift'] is True: mod = 1
            else:
                mod = 0

            ky.append(pylink.KeyInput(k, mod))

        return ky


    def exit_image_display(self):
        '''Clear the camera image'''

        self.clear_cal_display()
        self._display.flip()

    def alert_printf(self, msg):
        '''Print error messages.'''

        print("Error: " + msg)


    def setup_image_display(self, width, height):
        ''' set up the camera image

        return 1 to show high-resolution camera images'''

        self.last_mouse_state = -1
        self._size = (width, height)

        return 1


    def image_title(self, text):
        '''Draw title text below the camera image'''

        self._title.text = text


    def draw_image_line(self, width, line, totlines, buff):
        '''Display image pixel by pixel, line by line'''
```

```
        for i in range(width):
            try:
                self._imagebuffer.append(self._pal[buff[i]])
            except:
                pass

        if line == totlines:
            bufferv = self._imagebuffer.tostring()
            img = Image.frombytes("RGBX", (width, totlines), bufferv)
            self._img = ImageDraw.Draw(img)
            # draw the cross hairs
            self.draw_cross_hair()
            # scale the camera image
            self.imgResize = img.resize((width*2, totlines*2))
            cam_img = visual.ImageStim(self._display,
                                       image=self.imgResize,
                                       units='pix')
            cam_img.draw()
            # draw the camera image title
            self._title.pos = (0, - totlines - self._msgHeight)
            self._title.draw()
            self._display.flip()

            # clear the camera image buffer
            self._imagebuffer = array.array('I')

    def set_image_palette(self, r, g, b):
        '''Given a set of RGB colors, create a list of 24bit numbers
        representing the color palette.
        For instance, RGB of (1,64,127) would be saved as 82047,
        or 00000001 01000000 011111111'''

        self._imagebuffer = array.array('I')

        sz = len(r)
        i = 0
        self._pal = []
        while i < sz:
            rf = int(b[i])
            gf = int(g[i])
            bf = int(r[i])
            self._pal.append((rf << 16) | (gf << 8) | (bf))
            i = i+1
```

We did not discuss object-oriented programming in Chapter 1. Object-oriented programming is usually unnecessary for experimental tasks that typically involve only a couple hundred lines of code, but it is indispensable for large-scale applications. There is no need to delve into the details, but it is helpful to understand the concepts of *class* and *object*. An object is an instance of a class. For instance, the computer I am using to write up this book is an instance of the "laptop" class. This computer is classified as a "laptop" because it has the properties (e.g., small form factor, built-in keyboard and screen) and functions (e.g., computation, entertainment) that define a laptop. Once we have instantiated a concrete object of a class, we can retrieve or change its properties and request it to perform tasks for us. The tasks that an object can perform are known as "methods." More importantly, we can create a new class (e.g., "MacBook") based on a parent class ("laptop"), so the new class "MacBook" would inherit all the properties and methods of the "laptop" class.

The *EyeLinkCoreGrahicsPsychoPy* class defined in the custom CoreGraphics library is a child class, which inherits all the properties and methods of the Pylink *EyeLinkCustomDisplay* class. In this library, we overwrite the various methods (e.g., *draw_cal_target*) inherited from the *EyeLinkCustomDisplay* class with custom methods that work in PsychoPy. In this way the calibration routines will work smoothly in tasks created with PsychoPy.

```
class EyeLinkCoreGraphicsPsychoPy(pylink.EyeLinkCustomDisplay):
```

The first function *__init__(self, win, tracker)* is executed when an instance of the *EyeLinkCoreGraphicsPsychoPy* class is initialized. Here, we need to pass two arguments, i.e., the tracker connection we have established (*tracker*) and the window we would like to use for calibration (*win*). In this function, we also set a few properties, like the calibration screen background color and the warning beeps.

```
class EyeLinkCoreGraphicsPsychoPy(pylink.EyeLinkCustomDisplay):
    def __init__(self, tracker, win):
        '''Initialize

        tracker: an EyeLink instance (connection)
        win: the PsychoPy window we use for calibration'''

        pylink.EyeLinkCustomDisplay.__init__(self)

        # background and target color
        self._backgroundColor = win.color
        self._foregroundColor = 'black'

        # window to use for calibration
        self._display = win
```

In the CoreGraphics library, we define five groups of custom functions for calibration screen manipulation, calibration target drawing, warning beep playback, mouse and keyboard event handling, and camera image drawing.

```
def getColorFromIndex(self, colorindex):
    '''Retrieve colors for camera image elements, e.g., crosshair'''

def draw_line(self, x1, y1, x2, y2, colorindex):
    '''Draw a line

def draw_lozenge(self, x, y, width, height, colorindex):
    ''' draw a lozenge to show the defined search limits
```

For the calibration screen, we define the *setup_cal_display()*, *exit_cal_display()*, and *clear_cal_display()* functions. These functions are all about clearing up the calibration screen for additional drawings.

```
def setup_cal_display(self):
    '''Set up the calibration display '''

def clear_cal_display(self):
    '''Clear the calibration display'''

def exit_cal_display(self):
    '''Exit the calibration/validation routine'''
```

For calibration target drawing, the library has two functions, *erase_cal_target*() and *draw_cal_target()*. These functions will be called by the EyeLink API when it is time to draw or erase a calibration target on the screen.

```
def erase_cal_target(self):
    '''Erase the calibration/validation & drift-check target'''

def draw_cal_target(self, x, y):
    '''Draw the calibration/validation & drift-check target'''
```

The *play_beep()* function will play a warning beep to let the experimenter know that the calibration is good or not, etc.

```
def play_beep(self, beepid):
    '''Play warning beeps if being requested'''
```

The *getColorFromIndex()* function retrieves the color we use to draw the various graphics elements, for instance, the crosshair marking the pupil and CR center. The *draw_line()*, *draw_lozenge()* functions are called when we need to draw the crosses, the search limits, etc. on top of the camera image (see Fig. 7.4).

The CoreGraphics library also allows users to use the mouse to select which eye to track on the camera image and to press hotkeys like C to control the Host PC calibration routine. For this purpose, we define two functions to register keyboard and mouse events, *get_mouse_state()* and *get_input_key()*.

```
def get_mouse_state(self):
    '''Get the current mouse position and status'''

def get_input_key(self):
    '''This function is repeatedly pooled to check keyboard events'''
```

To show or hide the camera image on the Display PC, we need to define two functions, *setup_image_display()* and *exit_image_display()*. If the *setup_image_display()* function returns 1, the EyeLink API would request the Host PC to transfer high-resolution camera image (384 × 320 pixels).

```
def exit_image_display(self):
    '''Exit the camera image display'''

def setup_image_display(self, width, height):
    ''' set up the camera image

    return 1 to show high-resolution camera images'''

    self.last_mouse_state = -1
    self._size = (width, height)

    return 1
```

The *image_title()* function helps show the pupil and CR thresholds and the target distance (when tracking in Remote Mode) below the camera image on the Display PC (see Fig. 7.4). The Host PC will pass this info as a string through the Ethernet link, and the CoreGraphics library is responsible for showing this info to the experimenter.

```
def image_title(self, text):
    '''Show pupil CR thresholds below the image'''
```

We can request the Host PC to transfer the camera image to show on the Display PC. Upon request, the Host PC first converts the camera image pixels into an array of color palette indices, with each index representing a predefined color in a color palette. Once the Display PC has received the pixel indices, the EyeLink API uses the *draw_image_line()* function to reconstruct the image line by line, based on a predefined color palette, and show the image on the screen when reaching the last line of a camera image (i.e., when *line* == *totlines*). The image lines (rows of pixels) are buffered in an array (*self._imagebuffer*). The Image module of the Python Image Library (PIL, a dependency of PsychoPy) is used to convert this buffered array into an image; we then draw the crosshairs and search limits, etc. on the image, resize it, and show it on the PsychoPy window.

```
def draw_image_line(self, width, line, totlines, buff):
    '''Draw the camera image'''
```

As noted, the Host PC will send over the color palette indices for the camera image pixels. The Host PC also sends over the color palette that we can use to figure out the RGB values of each pixel. We store this palette into a global variable *self._pal*. This is done by the *set_image_palette()* method. When we read the image lines in the above *draw_image_line()* method, we extract the RGB values of each pixel based on its color palette index, *self._pal[buff[i]]*.

```
def set_image_palette(self, r, g, b):
    '''Get the color palette for the camera image'''
```

The crucial concept that users need to grasp here is that a custom CoreGraphics library is just a set of functions recognizable to Pylink. When Pylink is requested to perform a calibration-related task (e.g., draw the calibration target), the corresponding functions defined in the CoreGraphics library are executed. You can implement your library in Pygame or any other programming tools or libraries (e.g.,

OpenSesame). This is, however, unnecessary in most scenarios; such libraries are included in the EyeLink Developer's Kit or available from the SR Research Support website.

We have covered the frequently used functions of the Pylink library in Chaps. 4, 5, and 6. In this chapter, we further discussed three advanced features of the Pylink library. In the next and final chapter, we will switch gear and discuss eye movement data analysis and visualization.

Chapter 8
Eye Movement Data Analysis and Visualization

Contents

The analysis and visualization of eye movement data is a topic that deserves a dedicated book. There are lots of tools that can be used to analyze or visualize eye movement data. The EyeLink Developer's Kit comes with a set of libraries (e.g., the EDF access API) and tools that can be used to process the eye movement data. This chapter will give an overview of the format of the eye movement data stored in the EyeLink EDF data files and share a few techniques about data extraction and visualization. Manually extracting data from ASC files is usually unnecessary if you have a copy of the Data Viewer software, which helps to output hundreds of dependent measures. Nevertheless, the tips illustrated here can be useful if the required analysis is a relatively simple one, such as extracting the saccadic response times in an anti-/pro-saccade task, or if the analysis protocol is unavailable in Data Viewer, such as the calculation of saccade trajectory curvature.

Z. Wang, *Eye-Tracking with Python and Pylink*,
https://doi.org/10.1007/978-3-030-82635-2_8

EyeLink EDF Data File

EDF stands for "EyeLink Data Format." Unlike some other eye trackers that output data in plain text format, EyeLink data is saved in binary form to save disk space and, more importantly, to protect data integrity. The data contained in the EDF data files are accessible through the EDF Access API. Depending on the experimental setup, the EDF data files usually contain samples, online parsed eye events (e.g., saccades and fixations), messages, and additional information, like recording parameters. Here I will briefly discuss the format of each of these data types. For detailed information, please refer to the EyeLink 1000 Plus User Manual.

Samples

The EyeLink tracker images the eyes as fast as 2000 times a second. The Host PC will process every eye image to extract the pupil and CR centers. Sample data is associated with each of the eye images. It contains a timestamp at which the camera captured the eye image, the eye position data extracted from the eye image, the pupil size, and I/O events (e.g., parallel port INPUT). The EyeLink tracker records eye position data in multiple reference frames, i.e., PUPIL, HREF, and GAZE. PUPIL data is based on the raw coordinates on the camera sensor. The raw PUPIL data is very rarely required but can be useful for implementing custom calibration models (either online or offline). HREF is head-referenced position data, which measures eye rotation in the socket relative to the head. This type of data can be helpful in physiological studies that examine the actual eye movement velocity, etc. GAZE data is the most commonly used type of eye position data; it is the gaze position on a calibrated plane—typically, a computer screen. With a screen-based study, GAZE data represents the screen pixel coordinates that the eyes are looking at.

In addition to eye position data, the EyeLink sample data also contain information about pupil size. Pupil size data reflects the number of thresholded camera sensor pixels that the tracker recognizes as the pupil (multiplied by a scaling factor). Pupil size in these "arbitrary units" is rarely (if possible) directly compared between subjects, as the values will reflect differences in the eye to camera distance and camera angle. In a typical pupillometry task, researchers will use a "baseline" period and derive a "percentage change" measure for each experimental condition (see Mathôt et al. 2018, for a discussion on sensible baseline corrections of pupil data). The pupil size data reported by the EyeLink tracker is on a ratio scale, which allows the derivation of such a percentage measure. To convert the pupil size data to real-world units (e.g., millimeters), an artificial eye of known size (or a black dot printed on paper) can be recorded to derive a conversion factor. Detailed instructions are available on the SR Research Support website.

As most researchers use a flat screen to present stimuli, bear in mind that the number of pixels that correspond to 1 degree of visual angle will change with gaze position. The number of pixels that correspond to 1 degree of visual angle is known as "resolution" in EyeLink terminology. The resolution is the lowest when the eyes are looking at the screen center; the resolution is the highest when looking at the screen corners. The tracker always records resolution as part of the sample data, i.e., the number of pixels in the horizontal and vertical axes that equate to a single degree of visual angle at the current location of gaze.

The EDF Access API allows users to output some or all of the sample data described above. The format of a sample data line depends on the choice of the user. With resolution data, the columns of the data lines (from a monocular recording) contain the timestamp of the samples, followed by gaze position, pupil size, and resolution data. The sample data lines end with data flags. In the data lines below, "..." (three dots) mean that everything went well. If something went wrong, e.g., CR tracking is lost, letters will replace one or more of the trailing dots.

```
94711501    647.0    423.9    120.0    42.90    47.30    ...
94711502    647.9    425.0    120.0    42.90    47.40    ...
94711503    648.8    425.8    121.0    42.90    47.40    ...
94711504    649.7    426.7    121.0    42.90    47.40    ...
94711505    650.3    426.8    121.0    42.90    47.40    ...
94711506    650.7    426.3    121.0    42.90    47.40    ...
94711507    650.9    425.7    120.0    42.90    47.30    ...
94711508    650.8    425.0    120.0    42.90    47.30    ...
```

Sample data is typically most useful when you need to examine the temporal dynamics of eye movements, e.g., the probability of gaze in an interest area over time, the curvature of a saccade trajectory, or the pupil size changes following a critical experimental manipulation. It is also crucial for the time course (binning) analyses required when analyzing data from a Visual World paradigm.

Events

Parsing sample data into saccades and fixations is the focus of considerable research and can vary across research fields and as a function of hardware parameters such as sampling speed (see Hessels et al. 2018, for a recent survey among researchers). Generally speaking, the algorithms used to parse gaze data into saccades and fixations can be classified as either online (e.g., applied during recording) or offline (applied after the data has been recorded). Parsing algorithms can also be broadly divided into those based on gaze position dispersion and those based on eye movement velocity (Salvucci and Goldberg 2000). The algorithms used by the EyeLink trackers are velocity-based and applied online during recording. As such, they can be used to detect saccades and fixations with a reasonably short delay reliably. The online parsed eye events are typically saved in the EDF data files, together with the sample data. The data format for eye events depends on the type of event. For the start of an event, usually, there is only a timestamp in the data line. For instance, the following two data lines mark the start of a new fixation (SFIX) for both eyes.

```
SFIX L   111328822
SFIX R   111328822
```

For the end of an event, summary data is available, which starts with a keyword defining the end event type (e.g., ESACC for the termination of a saccade). The keyword is followed by a single-letter code indicating which eye the data refers to, timestamps indicating the start and end times of the event, and additional information (e.g., saccade duration, amplitude, etc.). The example data lines below include the start and end timestamps, saccade duration, start position in X and Y directions, end position in X and Y directions, amplitude, and peak velocity. The exact format of an event data line depends on the event type and the options you selected when retrieving data from the EDF files. Detailed information about all possible data line formats is available in the EyeLink 1000 Plus User Manual.

```
ESACC L 111329254 111329274 22 1650.2 428.0 1643.1 479.6 1.13 66
ESACC R 111329256 111329274 20 1657.6 425.5 1652.1 474.0 1.06 70
```

Events are indispensable in calculating frequently used dependent measures like fixation count, regressions, etc.

Other Useful Information in the EDF Data File

In addition to the eye movement data itself, the EDF data file also contains additional information indispensable for data analysis. This additional information includes messages, button events, and recording parameters.

Messages are vital for data analysis. Message data lines start with the keyword "MSG," followed by a timestamp, an optional time offset, and the message itself. An optional offset is occasionally used if a message cannot be sent in time, e.g., -13 in the data line shown below.

```
MSG    111321680 -13 DISPLAY ON
```

The messages are important for two reasons.

a) Messages mark the occurrence (onset time) of critical trial events, e.g., when the stimulus was drawn to the screen. Without messages, it would be difficult to tell the timing of different trial events in a study.
b) Special messages can be used to store the information required to properly analyze and visualize eye movement data in Data Viewer. These messages usually start with "!V", followed by a command (e.g., "IMGLOAD CENTER") and additional information (see Chap. 6).

The beginning and end of a data recording block are marked with the keywords "START" and "END." The recording parameters are stored in a data line that starts with the "RECCFG" keyword (i.e., recording configuration). The pupil and CR thresholds in a recording session are stored in a data line that starts with the keyword "THRESHOLDS." For more information, please refer to the EyeLink user manual.

EDF Converter

SR Research has provided an EDF converter for users to extract eye movement data from the EDF data files. The extracted data are stored in plain text files, with an extension of .ASC (see Fig. 8.1). Python can be used to read in the ASCII file line by line.

The EDF converter has a very intuitive user interface. The format of the output files can be configured via the Preference menu. In Fig. 8.2, the "Output Resolution Data" option is checked, so the sample data lines in the converted ASC file will have two extra columns specifying how many pixels correspond to 1 degree of visual angle in both horizontal and vertical directions.

The EyeLink Developer's Kit also comes with a command-line version of the EDF converter. On macOS, you can find this command-line tool in the following folder.

```
/Applications/EyeLink/EDF_Access_API/Example
```

You may create a symbolic link in /usr/local/bin to make "edf2asc" accessible from a terminal.

```
sudo ln -s /Applications/EyeLink/EDF_Access_API/Example/edf2asc /usr/
local/bin/edf2asc
```

On Windows, you can find the command-line tool in the EDF Access API folder as well.

```
C:\Program Files (x86)\SR Research\EyeLink\EDF_Access_API\Example
```

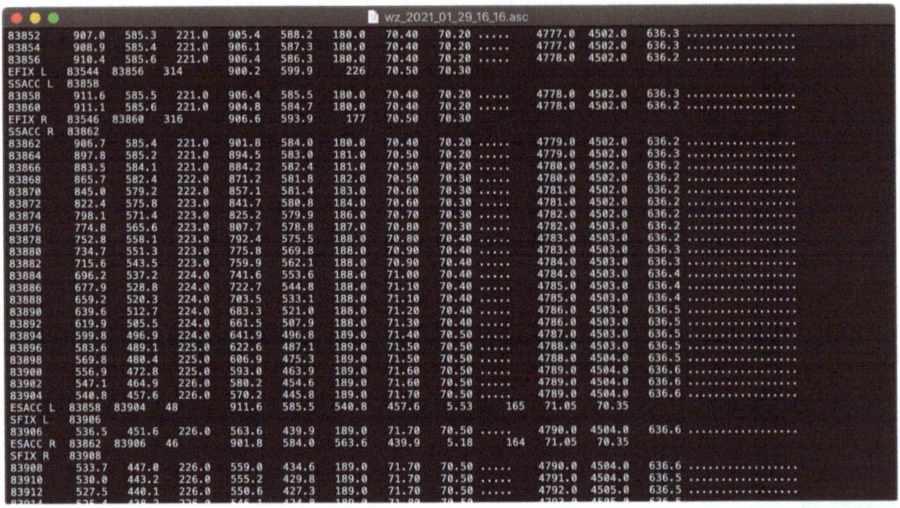

Fig. 8.1 An example ASC data file generated by the EDF converter

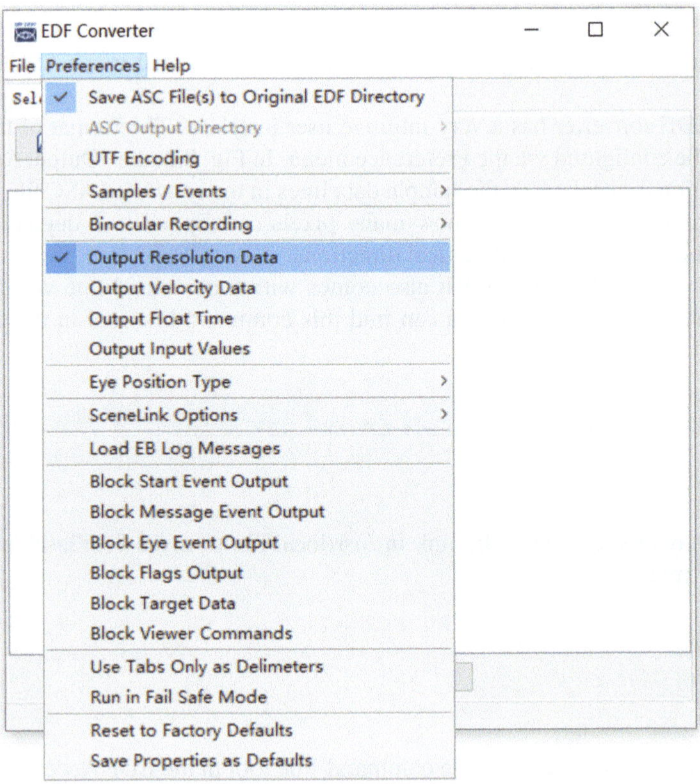

Fig. 8.2 The EDF converter provided by SR Research allows users to export different data types

To make the *edf2asc* command-line tool system-wide accessible, go to "My computer -> properties -> advanced -> environment variables -> Path," and add the above directory.

Once the *edf2asc* command-line tool is system-wide accessible, you can use the following command to convert EDF data files into ASC files.

```
edf2asc [options] <input .edf file>
```

There are lots of options you can set when using this command, which mirror the preferences available in the GUI apps (Fig. 8.2). For instance, the option "-p <path>" specifies where to save the converted ASC files. The output file name can be specified. If it is not specified, the EDF converter will use the input file name plus the ".asc" extension to save the converted files. Entering "edf2asc" without any options will print all possible options in the command line.

You can use *os.system*() to execute the "edf2asc" command in a Python script to convert EDF data files into ASC format. For instance, the following code will convert all the EDF data files in the current working directory. Here, the command passed to *os.system()* is "edf2asc -y *.edf." The option "-y" requests *edf2asc* to convert the EDF data file without user confirmation. The * in the command is a wildcard, i.e., a character that represents zero or more characters in a string. The string "*.edf" will give us all files that end with a ".edf" extension in the current working directory.

```
import os
os.system('edf2asc -y *.edf')
```

Extract Data from the ASC Files

The converted ASC files are plain text files. You can use any programming language to extract data from them. A typical workflow is to read an ASC file line by line to scan for useful information. In the ASC files, lines that start with "*" are preamble texts (file headers), and those that start with "#" , ";" , or "/" are comments. These lines are safe to ignore. Sample data lines always start with a timestamp, i.e., numbers constitute 0–9, whereas the data lines for eye events, message, etc. always start with capital letters (e.g., keywords "SSACC" and "ESACC"). You may also see dots (".") that represent missing values in the sample data lines during which blinks occur.

Here I will present two techniques for extracting data from the converted ASC files. Before you go down this road, you may also want to look at other freely available tools that might be useful for your data analysis aims, such as the "Python EyeLinkParser"[1] by Sebastiaan Mathôt and "pypillometry"[2] by Matthias Mittner.

Parse ASC Files with the String Function split()

In this example, we will plot the scanpath over a background image. The sample eye movement data were recorded in a free-viewing task (see Chap. 4) where the participants passively viewed a picture on each trial. The EDF data file is stored in a folder named "freeview," along with the images presented in the task (see below for the folder structure). Let us assume that we have converted the EDF file (freeview.edf) and saved the resulting ASC file in the same folder as the EDF data file. The following short script will read in the data lines in the ASC file, look for fixation events and look for messages that contain information about the image presented on each trial, and finally draw the scanpath over the relevant image.

```
|__ freeview
|       |__ freeeview.edf
|       |__ freeview.asc
|       |__ images
|               |__ quebec.jpg
|               |__ woods.jpg
|
|__ parse_ASC_4scanpath.py
```

The scanpath of an experimental trial is shown in Fig. 8.3, and the short script used to generate this figure is presented below.

To keep the script short and simple, we use the *Image* and *ImageDraw* modules from the PIL library to manipulate the background image and to draw the scanpath. We first import these modules, of course. We then open the converted ASC file and read the contents of the ASC file line by line in a for loop. We call the string method *rstrip*() to strip trailing space in the data line, if there were any. We then call the string method *split*() to convert the data line into a Python list.

```
tmp_data = line.rstrip().split()
```

[1] https://github.com/smathot/python-eyelinkparser

[2] https://ihrke.github.io/pypillometry/html/index.html

Fig. 8.3 A scanpath constructed from an ASC file

```
#!/usr/bin/env python3
#
# Filename: parse_ASC_4scanpath.py
# Author: Zhiguo Wang
# Date: 5/25/2021
#
# Description:
# Parse an ASC file to extract fixations, then plot the scanpath.

import os
from PIL import Image, ImageDraw
from math import sqrt

# Open the converted ASC file
asc = open(os.path.join('freeview', 'freeview.asc'))

new_trial = False
trial = 0
for line in asc:
    # Convert the current data line into a list
    tmp_data = line.rstrip().split()

    # Get the correct screen resolution from the DISPLAY_COORDS message
    # MSG    4302897 DISPLAY_COORDS 0 0 1279 799
```

```
if 'DISPLAY_COORDS' in line:
    scn_w = int(tmp_data[-2]) + 1
    scn_h = int(tmp_data[-1]) + 1

# Look for the message marking image onset
if 'image_onset' in line:
    new_trial = True
    trial += 1
    print(f'Processing trial # {trial} ...')

    # Store the position and duration of all fixations in lists
    fix_coords = []
    fix_duration = []

if new_trial:
    # Path to the background image
    # MSG    3558923 !V IMGLOAD FILL images/woods.jpg
    if 'IMGLOAD' in line:
        bg_image = tmp_data[-1]

    # Retrieve the coordinates and duration of all fixations
    # EFIX R 80790054 80790349 296 981.3 554.5 936 63.50 63.50
    if 'EFIX' in line:
        duration, x, y = [int(float(x)) for x in tmp_data[4:7]]
        fix_coords.append((x, y))
        fix_duration.append(duration)

# Look for the message marking image offset, draw the scanpath
if 'image_offset' in line:
    # Open the image and resize it to fill up the screen
    img = os.path.join('freeview', bg_image)
    pic = Image.open(img).resize((scn_w, scn_h))

    # Create an ImageDraw object
    draw = ImageDraw.Draw(pic)

    # Draw the scanpath
    draw.line(fix_coords, fill=(0, 0, 255), width=2)

    # Draw circles to represent the fixations, the diameter reflects
    # the fixation duration, scaled to its maximum
    for i, d in enumerate(fix_duration):
        sz = sqrt(d / max(fix_duration) * 256)
        gx, gy = fix_coords[i]
        draw.ellipse([gx-sz, gy-sz, gx+sz, gy+sz],
                     fill=(255, 255, 0), outline=(0, 0, 255))
```

```
        # Save the scanpath for each trial
        pic.save(f'scanpath_trial_{trial}.png', 'PNG')

# Close the ASC file
asc.close()
```

For instance, for the following data line, the list returned by the above command is ['MSG', 4302897, DISPLAY_COORDS', '0.00', '0.00', '1279.00', '799.00']. The DISPLAY_COORDS data line records the pixel coordinates of the top-left and bottom-right corners of the display. When the screen resolution is 1280 x 800 pixels, the coordinates of the top-left corner are (0, 0), and that for the bottom-right corner is the screen resolution minus 1, i.e., (1279, 799).

```
MSG    4302897 DISPLAY_COORDS 0 0 1279 799
```

To properly overlay the gaze on the screen, we need to know the screen size in pixels. We look for this info from the "DISPLAY_COORDS" data line (see above), which was sent to the tracker every time we start a new recording block. Note that *tmp_data[-1]* will give us the last item in *tmp_data*, and *tmp_data[-2]* will give us the second last item in *tmp_data*.

```
        scn_w = int(tmp_data[-2]) + 1
        scn_h = int(tmp_data[-1]) + 1
```

The example EDF data file contains "image_onset" messages marking the time at which an image appeared on the screen. We set the *new_trial* variable to *True* if we encounter a data line containing this message. Note that we also created two empty lists to store the fixation coordinates and durations, i.e., *fix_coord = []* and *fix_duration = []*. We use these lists to store the position and duration of all fixations that occurred during image presentation.

```
    # Look for the message marking image onset
    if 'image_onset' in line:
        new_trial = True
        trial += 1
        print(f'Processing trial # {trial} ...')

        # Store the position and duration of all fixations in lists
        fix_coords = []
        fix_duration = []
```

In the free-viewing task, the experimental script logged an "IMGLOAD" command to the EDF data file to point to the images presented during testing.

```
MSG    3558923 !V IMGLOAD FILL images/woods.jpg
```

While looping over the data lines from the converted ASC file, we look for the "IMGLOAD" keyword and extract the path to the background image.

```
        if 'IMGLOAD' in line:
            bg_image = tmp_data[-1]
```

From data lines that start with the "EFIX" keyword, we extract the fixation coordinates and duration; then append them to the *fix_coords* and *fix_duration* lists. Later on, we can use the list *fix_coords* to construct the scanpath.

```
        # Retrieve the coordinates and duration of all fixations
        # EFIX R 80790054 80790349 296 981.3 554.5 936 63.50 63.50
        if 'EFIX' in line:
            duration, x, y = [int(float(x)) for x in tmp_data[4:7]]
            fix_coords.append((x, y))
            fix_duration.append(duration)
```

At the end of each trial, the experimental script cleared the screen and sent an "image_offset" message to the tracker to log this event. When we see this message in the ASC file, we open the background image file and use it as a canvas for drawing. Then, we draw the scanpath and save the resulting canvas into a PNG file. In the resulting scanpath plot, fixations are represented by dots of varying sizes (reflecting the duration of the fixations). [3]

```
    # Look for the message marking image offset, draw the scanpath
    if 'image_offset' in line:
        # Open the image and resize it to fill up the screen
        img = os.path.join('freeview', bg_image)
        pic = Image.open(img).resize((scn_w, scn_h))

        # Create an ImageDraw object
        draw = ImageDraw.Draw(pic)
```

[3] To keep the script simple, we stop looking for the EFIX event when encountering the "image_offset" message. So, the script may omit the last fixation if it overlaps with this message.

```
        # Draw the scanpath
        draw.line(fix_coords, fill=(0, 0, 255), width=2)

        # Draw circles to represent the fixations, the diameter reflects
        # the fixation duration, scaled to its maximum
        for i, d in enumerate(fix_duration):
            sz = sqrt(d / max(fix_duration) * 256)
            gx, gy = fix_coords[i]
            draw.ellipse([gx-sz, gy-sz, gx+sz, gy+sz],
                         fill=(255, 255, 0), outline=(0, 0, 255))

        # Save the scanpath for each trial
        pic.save(f'scanpath_trial_{trial}.png', 'PNG')
```

We close the ASC file after finishing processing all experimental trials.

```
# close the ASC file
asc.close()
```

Parse ASC Files with Regular Expressions

In the previous example, we illustrated how to use the string method *split*() to extract eye movement data and event/message information from an ASC file. An alternative approach is to use "regular expressions," though the underlying principles are essentially the same. We read the ASC file line by line, scan for keywords, and grab the data of interest based on the data line format. In the short script below, we extract all the fixation and saccade END events and put them into *Pandas* data frames to use for further analysis. Pandas is a Python library for data analysis and manipulation; a data frame in Pandas is a 2D data table of rows and columns, much like a spreadsheet.[4]

```
#!/usr/bin/env python3
#
# Filename: parse_ASC_re.py
# Author: Zhiguo Wang
# Date: 5/25/2021
#
# Description:
# Parse the ASC file with regular expressions (re).
```

[4] https://pandas.pydata.org/

```
import os
import re
import pandas as pd

# Open the converted ASC file
asc = open(os.path.join('freeview', 'freeview.asc'))

efix = []  # fixation end events
esac = []  # saccade end events
for line in asc:
    # Extract all numbers and put them in a list
    tmp_data = [float(x) for x in re.findall(r'-?\d+\.?\d*', line)]

    # retrieve events parsed from the right eye recording
    if re.search('^EFIX R', line):
        efix.append(tmp_data)
    elif re.search('^ESACC R', line):
        esac.append(tmp_data)
    else:
        pass

# Put the extracted data into pandas data frames
# EFIX R 80790054 80790349 296 981.3 554.5 936
efix_colname = ['startT', 'endT', 'duration', 'avgX', 'avgY', 'avgPupil']
efixFRM = pd.DataFrame(efix, columns=efix_colname)
# ESACC R 80790350 80790372 23 982.6 551.8 864.9 587.9 1.94 151
esac_colname = ['startT', 'endT', 'duration', 'startX', 'startY',
                'endX', 'endY', 'amplitude', 'peakVel']
esacFRM = pd.DataFrame(esac, columns=esac_colname)

# Close the ASC file
asc.close()
```

In this example script, we open the ASC file and create two empty lists to store the fixation and saccade END events. Then, we examine the data lines in a for loop. When scanning the data lines, we first extract the numbers, if there are any, with *re.findall()* function from the standard Python module "re" (regular expression). Regular expressions are a sophisticated matter, and interested users can consult the tutorials from the official Python documentation.[5] I will nevertheless explain the regular expressions used in the present example, notably the following commands.

```
re.findall(r'-?\d+\.?\d*', line)
```

[5] https://docs.python.org/3/library/re.html

The first argument we pass to *re.findall()* is a "pattern" constructed with regular expressions. Here, the question mark ("?") means that the immediately preceding element (on the left) should repeat 0 or 1 times. So, "-?" allows us to look for both positive (without "-") or negative (with "-") values. "\d" represents Unicode decimal digits [0–9] and other digit characters. The plus sign "+" requires the preceding regular expression to repeat at least once, so "\d+" will give us a string of numbers. The "+" sign is followed by "\.", to include a decimal point (".") in the pattern. The escape character "\" is required because "." can represent anything in regular expressions. We need to "escape" this feature, so a dot is recognized as a regular decimal point. The dot is followed by another "?" to make the decimal point optional in the pattern. Then, another "\d" followed by an asterisk ("*"), which means that zero or more digits should follow the decimal point. The *re.findall()* call will return a numeric string, which we convert to a float number for further analysis.

The regular expression we used to identify the different types of data lines was caret ("^"), which examines only the start of a string to look for matches. Lines that start with "EFIX" are data lines for "fixation end" events, and those that start with "ESACC" are data lines for "saccade end" events.

```
# Retrieve events parsed from the right eye recording
if re.search('^EFIX R', line):
    efix.append(tmp_data)
elif re.search('^ESACC R', line):
    esac.append(tmp_data)
else:
    pass
```

The last job of this short script is to put the extracted fixation and saccade END events into Pandas data frames that we can manipulate in further analysis. Column headers are added to make the data frame columns easier to interpret and work with. For a fixation end event, the columns are fixation start time, end time, fixation duration, average gaze position in X and Y directions, and average pupil size.

```
>>> efixFRM
       startT        endT  duration   avgX   avgY  avgPupil
0  94711370.0  94711843.0     474.0  645.5  421.4     119.0
1  94711859.0  94711881.0      23.0  642.7  421.4     118.0
2  94711927.0  94712064.0     138.0  449.2  279.3     117.0
3  94712111.0  94712339.0     229.0  664.9  408.2     111.0
4  94712362.0  94712551.0     190.0  692.2  448.6     107.0
5  94712599.0  94712899.0     301.0  557.1  231.0     115.0
6  94712946.0  94713129.0     184.0  321.6  191.5     113.0
```

Data Visualization

The promise of revealing the observer's gaze behavior makes data visualization an essential aspect of eye movement data analysis. There are various data visualization approaches, ranging from simple scanpath and "heatmap" for revealing the overall visual selection process to complex chord diagrams showing the transition of gaze between objects or interest areas. Although data visualization cannot substitute rigorous statistical analysis, it is an effective way to convey research results if used properly. In the following section, we demonstrate how to implement a few common visualization techniques in Python. For a thorough survey of the various eye movement visualization techniques that have been used by researchers, readers may find a recent literature review by Blascheck et al. (2017) an interesting read.

This section presents the visualization techniques based on the level of data abstraction (samples -> events -> interest areas).

Gaze Trace Plots

Gaze trace plots are frequently used to visualize sample data in commercial data analysis software and neurophysiological publications. This type of figure shows the gaze position in horizontal and vertical directions over time. We can use a gaze trace plot to visualize oculomotor responses to various experimental manipulations—for example, the gaze position changes following the onset of a peripheral visual target. Some commercial software (e.g., Data Viewer) also allows users to visually inspect the detected eye events (saccades and fixations) against the trace of gaze, so it is possible to examine how well the event parsing algorithm worked in detecting eye events.

There are many Python libraries that can be used to visualize the gaze trace over time (e.g., *Matplotlib,*[6] *Seaborn,*[7] and *Bokeh*[8]). The first step is to extract the sample data from the ASC file. Then, all you need is a line plot. Note that some sample data lines in the ASC file may contain missing values, for instance, due to blinks. By default, missing values are represented with a "." in the ASC files (see example data lines below).

```
80855874     1506.4      269.0       729.0    ...
80855875       .           .      0.0     ...
```

Please see below for a short script that plots the gaze trace from a single trial.

[6] https://matplotlib.org/

[7] https://seaborn.pydata.org/

[8] https://bokeh.org/

```
#!/usr/bin/env python3
#
# Filename: gaze_trace_plot.py
# Author: Zhiguo Wang
# Date: 5/25/2021
#
# Description:
# Extract the samples from an ASC file, then plot a gaze trace plot.

import os
import re
import numpy as np
import pandas as pd
import matplotlib.pyplot as plt

# Convert EDFs to ASC files with the edf2asc command-line tool
# If you run this script from IDLE on macOS, be sure to launch IDLE
# from the command-line (e.g., enter "idle3.6" in the terminal)
#
# Options for the command line "edf2asc" converter
#     -r, output right-eye data only
#     -y, overwrite ASC file if exists
cmd = 'edf2asc -r -y freeview/freeview.edf'
os.system(cmd)

# Open the converted ASC file
asc = open(os.path.join('freeview', 'freeview.asc'))

new_trial = False
trial_DFs = {}  # samples from all trials in a tuple
trial = 0
for line in asc:
    # Extract numerical values from the data line
    values = [float(x) for x in re.findall(r'-?\d+\.?\d*', line)]

    # Look for the message marking image onset
    if re.search('image_onset', line):
        new_trial = True
        trial += 1
        print(f'processing trial # {trial}...')

        # Store samples in lists (timestamp, x, y, pupil size)
        tmp_DF = []

    # A sample data line always starts with a numerical literal
    if new_trial and re.search('^\d', line):
        # 80855874     1506.4     269.0      729.0    ...
        # 80855875       .         .          0.0    ...
```

```
        if len(values) == 4:  # normal sample line
            tmp_DF.append(values)
        else:  # sample line with missing values (e.g., tracking loss)
            tmp_DF.append([values[0], np.nan, np.nan, np.nan])

    if re.search('image_offset', line):  # message marking image offset
        # Put samples in a pandas data frame and store it in trial_DFs
        colname = ['timestamp', 'gaze_x', 'gaze_y', 'pupil']
        trial_DFs[trial] = pd.DataFrame(tmp_DF, columns=colname)
        new_trial = False

# close the ASC file
asc.close()

# Plot the gaze trace and pupil size data from trial # 1
trial_DFs[1].plot(y=['gaze_x', 'gaze_y', 'pupil'])
plt.show()
```

At the beginning of the script, we imported a few Python modules for data manipulation (NumPy[9] and Pandas) and visualization (matplotlib.pyplot).

```
import numpy as np
import pandas as pd
import matplotlib.pyplot as plt
```

Instead of using the GUI version of the EDF converter, we call the command-line tool "edf2asc" to convert the EDF data files into ASC files. To call the *edf2asc* command, we import the standard Python module *os* and add the following lines to the script. The "-r" option helps to extract right-eye data only if the recording is binocular. The "-y" option for the *edf2asc* command will overwrite an ASC file if it exists. If you are running the script from IDLE on macOS, be sure to launch IDLE from a terminal.

```
cmd = 'edf2asc -r -y freeview/freeview.edf'
os.system(cmd)
```

[9] https://numpy.org/

We then open the ASC file and read and store the sample data from each experimental trial in a Pandas data frame. We put these data frames in a dictionary and label each data frame with the corresponding trial number, e.g., 1, 2, 3, 4, 5. The following pseudo-code illustrates why this method is useful for storing sample data from the experimental trials. To retrieve the Pandas data frame that stores the sample data from trial 1, call *trial_DFs[1]*.

```
trial_DFs = {1:dataFrame1, 2:dataFrame2, 3:dataFrame3}
```

In the example script, we examine the ASC file line by line with a for loop. For each of the data lines, we extract all numerical strings with a regular expression, *re.findall(r '-?\d+\.?\d*', line)*; then, we use the list comprehension trick to convert strings into numbers.

```
# Extract numerical values from the data line
values = [float(x) for x in re.findall(r'-?\d+\.?\d*', line)]
```

We are only interested in data recorded during image presentation. The "image_onset" message marks the onset of the image. Once we have detected this message in a data line, we create a list to store the samples.

```
# Look for the message marking image onset
if re.search('image_onset', line):
    new_trial = True
    trial += 1
    print(f'processing trial # {trial}...')

    # Store samples in lists (timestamp, x, y, pupil size)
    tmp_DF = []
```

Because we extracted right-eye data only from the EDF data file, a sample line by default contains four numbers (timestamp, gazeX, gazeY, pupil size). In contrast, missing sample data lines have only two numerical values (see the example data lines below).

```
80855874    1506.4    269.0    729.0    ...
80855875    .    .    0.0    ...
```

A sample data line always starts with a timestamp. We use the *re.search()* function to check if a data line represents sample data, *re.search('^\d,' line)*; if so, we further examine how many numbers are in this line. If there are missing values, we represent them with a NumPy constant, *np.nan*.

```
# A sample data line always starts with a numerical literal
if new_trial and re.search('^\d', line):
    # 80855874    1506.4    269.0   729.0    ...
    # 80855875      .   .        0.0   ...
    if len(values) == 4:  # normal sample line
        tmp_DF.append(values)
    else:  # sample line with missing values (e.g., tracking loss)
        tmp_DF.append([values[0], np.nan, np.nan, np.nan])
```

We then append the sample data, [timestamp, x, y, pupil size], to a list (*tmp_DF*). If the "image_offset" message is detected, we convert the sample data into the Pandas data frame and store it in the dictionary created outside the *for* loop (*trial_DFs*).

```
    if re.search('image_offset', line):  # message marking image offset
        # Put samples in a pandas data frame and store it in trial_DFs
        colname = ['timestamp', 'gaze_x', 'gaze_y', 'pupil']
        trial_DFs[trial] = pd.DataFrame(tmp_DF, columns=colname)
```

The resulting plot for trial 1 is presented in Fig. 8.4, which shows the x, y gaze position and pupil size over 13 seconds. Note a blink occurred toward the end of the image presentation.

```
# Plot the gaze trace and pupil size data from trial # 1
trial_DFs[1].plot(y=['gaze_x', 'gaze_y', 'pupil'])
plt.show()
```

Heatmap

A heatmap can also be referred to as an attention map, fixation map, or luminance map. It is a way of visualizing fixations over a background stimulus (usually a static image), and it helps to identify regions or objects that are of interest to the viewers. There are different types of heatmaps, depending on the data being used to construct the map, e.g., fixation duration, fixation count, relative fixation duration, or percentage of subjects. The nature of the data can potentially constrain the usefulness and interpretation of a heat map. For instance, to examine search efficiency, a heatmap

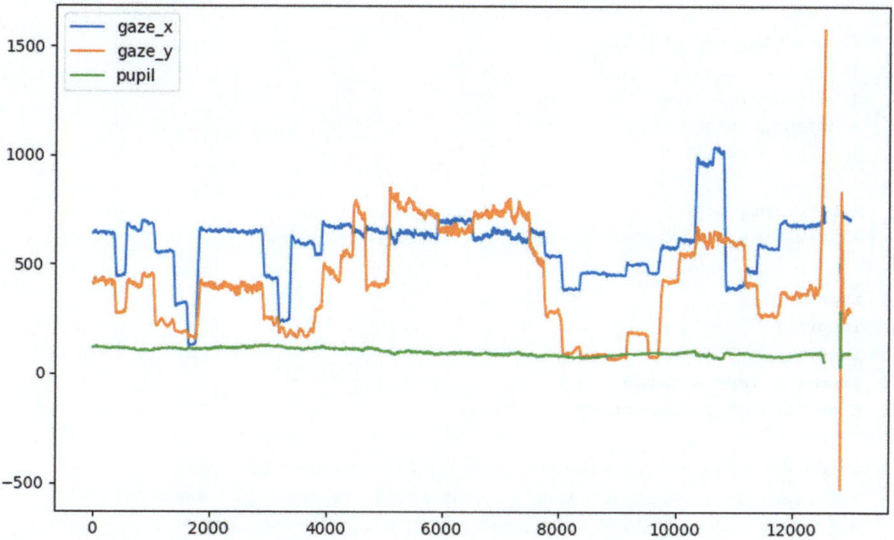

Fig. 8.4 A gaze trace plot created with the Pandas and Matplotlib libraries

based on fixation count (until a response is issued) may be more useful than a heatmap based on fixation duration. However, the fixation count itself will be affected by the total time the viewers spent examining the stimuli, leading to interpretation constraints. Another critical factor to consider is the criteria used to define "fixations," e.g., the minimal duration. In addition, because heatmaps aggregate fixations over time, temporal sequence information is lost. Please see Bojko (2009) for a discussion on the various problematic issues that can arise when using and interpreting heatmaps.

This section illustrates how to create a simple heatmap based on fixation duration. A duration-based heatmap is the summation of 2D Gaussians. The fixation duration determines the height of each Gaussian, and the fixation position determines the center of each Gaussian. There is no consensus over the spread of the 2D Gaussians. Generally speaking, it is constrained by the size of the most useful foveal visual field (~ 1 degree in radius) or the eye tracker's known tracking accuracy, usually in the 0.15–1.0-degree range. In either case, 1 degree of visual angle appears to be a safe choice. As noted, resolution data is available in the EyeLink EDF data files; so, it is easy to figure out how many pixels correspond to 1 degree of visual angle for each fixation. To include resolution data in the converted ASC file, the "*-res*" option is passed to the *edf2asc* command. The option "*-e*" requires *edf2asc* to extract eye events only, so the script scans fewer data lines.

```
#!/usr/bin/env python3
#
# Filename: heatmap_simple.py
# Author: Zhiguo Wang
# Date: 4/28/2021
#
# Description:
# Extract fixations from an ASC file to create a heatmap

import os
import re
import numpy as np
from PIL import Image
from matplotlib import cm

# Convert EDFs to ASC files with the edf2asc command-line tool
# If you run this script from IDLE on macOS, be sure to launch IDLE
# from the command-line (e.g., enter "idle3.6" in the terminal)
#
# Options for the command line "edf2asc" converter
#     -e, output event data only
#     -res, output resolution data if present
#     -y, overwrite ASC file if exists
cmd = 'edf2asc -e -res -y freeview/freeview.edf'
os.system(cmd)

# Open the converted ASC file
asc = open('freeview/freeview.asc', 'r')

# Transparency for the heatmap
alpha = 0.5

new_trial = False
trial = 0
for line in asc:
    # Extract all numerical values from the data line
    values = [float(x) for x in re.findall(r'-?\d+\.?\d*', line)]

    # Get the correct screen resolution from the GAZE_COORDS message
    # MSG    4302897 DISPLAY_COORDS 0 0 1279 799
    if re.search('DISPLAY_COORDS', line):
        scn_w = int(values[-2]) + 1
        scn_h = int(values[-1]) + 1
```

```
    # Look for the message marking image onset
    if re.search('image_onset', line):
        new_trial = True
        trial += 1
        print(f'processing trial # {trial}...')

        # Initialize the heatmap matrix
        w, h = np.meshgrid(np.linspace(0, scn_w, scn_w),
                           np.linspace(0, scn_h, scn_h))
        heatmap = np.exp(-(w**2 + h**2)) * 0

    if new_trial:
        if re.search('EFIX', line):
            # EFIX R 80790373 80790527 155 855.5 596.0 881 63.60 63.75
            start_t, end_t, duration, x, y, peakv, res_x, res_y = values

            # add the new fixation to the heatmap
            heatmap += duration * np.exp(-(1.0*(w - x)**2/(2*res_x**2) +
                                           1.0*(h - y)**2/(2*res_y**2)))

        # Path to the background image
        # MSG    3558923 !V IMGLOAD FILL images/woods.jpg
        if 'IMGLOAD' in line:
            bg_image = line.rstrip().split()[-1]

    # Look for the message marking image offset, create a heatmap
    if re.search('image_offset', line):
        # Open the image and resize it to fill up the screen
        img = os.path.join('freeview', bg_image)
        pic = Image.open(img).convert('RGBA').resize((scn_w, scn_h))

        # Apply a colormap (from the colormap library of MatplotLib)
        heatmap = np.uint8(cm.jet(heatmap/np.max(heatmap))*255)

        # blending
        heatmap = Image.fromarray(heatmap)
        blended = Image.blend(pic, heatmap, alpha)

        # Save the heatmap as a PNG file
        blended.save(f'heatmap_trial_{trial}.png', 'PNG')
        new_trial = False

# Close the ASC file
asc.close()
```

The script is much like the scanpath plot shown in Fig. 8.3, except that we first use a mesh grid to create the initial heatmap matrix when a new trial starts.

```
# Initialize the heatmap matrix
w, h = np.meshgrid(np.linspace(0, scn_w, scn_w),
                   np.linspace(0, scn_h, scn_h))
heatmap = np.exp(-(w**2 + h**2)) * 0
```

Every time we encounter a new fixation, we add a 2D Gaussian to the heatmap matrix, based on the current fixation position and duration. The spread of the 2D Gaussian is set to the resolution at the current gaze position (1°).

```
if re.search('EFIX', line):
    # EFIX R 80790373 80790527 155 855.5 596.0 881 63.60 63.75
    start_t, end_t, duration, x, y, peakv, res_x, res_y = values

    # add the new fixation to the heatmap
    heatmap += duration * np.exp(-(1.0*(w - x)**2/(2*res_x**2) +
                                   1.0*(h - y)**2/(2*res_y**2)))
```

At the end of a trial, we load a colormap from the Matplotlib library. Then, we use *Image.blend()* to blend the heatmap with the background image (see Fig. 8.5).

```
# Look for the message marking image offset, create a heatmap
if re.search('image_offset', line):
    # Open the image and resize it to fill up the screen
    img = os.path.join('freeview', bg_image)
    pic = Image.open(img).convert('RGBA').resize((scn_w, scn_h))

    # Apply a colormap (from the colormap library of MatplotLib)
    heatmap = np.uint8(cm.jet(heatmap/np.max(heatmap))*255)

    # blending
    heatmap = Image.fromarray(heatmap)
    blended = Image.blend(pic, heatmap, alpha)

    # Save the heatmap as a PNG file
    blended.save(f'heatmap_trial_{trial}.png', 'PNG')
    new_trial = False
```

Fig. 8.5 A heatmap created with the NumPy and PIL libraries

Scanpath

A scanpath is a sequence of alternating fixations and saccades, which can provide information about the visual search strategy (behavior) of the viewers. A scanpath typically constitutes circles (representing fixations) connected by straight lines (representing saccades). The diameter of the circles can be varied to represent the duration of fixations. In the simplest form of a scanpath plot, the fixation circles can be omitted, showing only the saccade path. We have presented an example script for drawing scanpaths in the previous section (see Fig. 8.3). Additional information, such as the fixations sequence, can be shown in the plot by labeling the fixations with indices. We can also use arrows to show the direction of saccades. We can, of course, also show the fixation duration with labels.

One big challenge for this visualization technique is cluttering. For recordings longer than a few seconds, it becomes challenging to determine the sequence of fixations and saccades. While scanpath visualizations can be useful representations of individual trials, comparing scanpaths across trials or conditions is a complex task. For various statistical approaches in comparing scanpaths, interested readers are referred to the paper by Anderson et al. (2015).

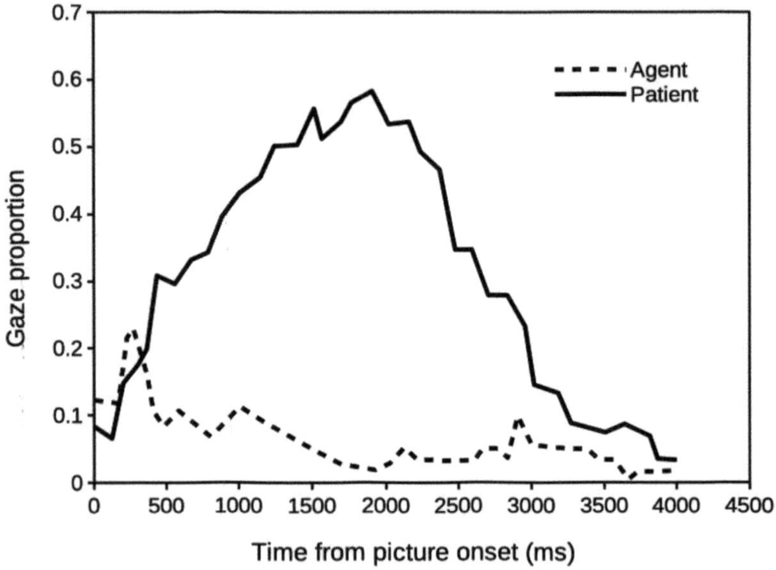

Fig. 8.6 A timeline plot showing the percentage of gaze in each interest area over time. (Redrawn from Griffin and Bock 2000)

Interest Area-Based Plots

Eye events abstract the raw sample data to a level that affords psychological meanings. One further step of abstraction is grouping eye events, notably fixations, based on the "semantics" in a scene to reveal which objects or regions attract the eyes. This type of analysis usually involves creating interest areas or "regions of interest," which can be used to generate dependent measures (e.g., dwell time and fixation count) for further statistical analysis.

Interest areas can also be useful for data visualization. One frequently seen visualization technique is a timeline plot constructed from raw sample data to reveal the temporal characteristic of decision processes. This visualization technique capitalizes on the fact that the eyes can only gaze at one location at a given moment. For instance, in a widely cited paper by Griffin and Bock (2000), participants described drawings with simple sentences like "The dog is chasing the mailman." In one of the tasks, the authors required the participant to tell who the "patient" (a person or thing being acted on in an event, e.g., "mailman") was while recording eye movements. The authors created a timeline plot based on the probability of looking at the agent ("dog") and the patient ("mailman") in continuous 4-ms time bins. The resulting plot revealed that people could identify the patient in about 288 ms following picture onset (i.e., gaze to the patient began to diverge from gaze to the agent; see Fig. 8.6).

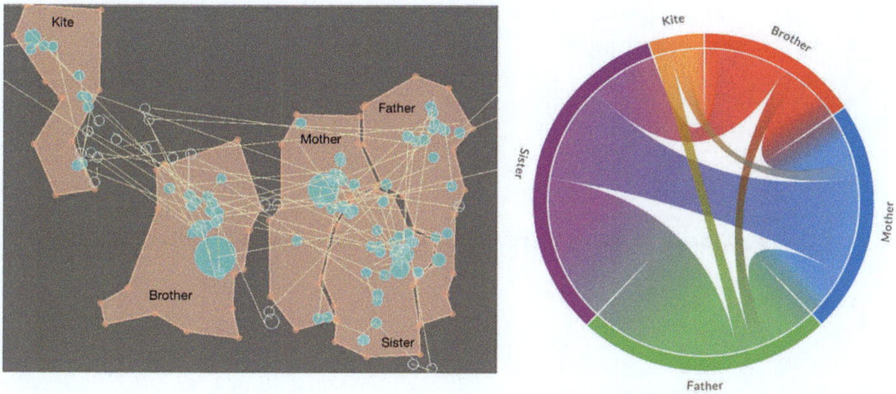

Fig. 8.7 A scanpath plot showing the fixation transitions between interest areas in a free-viewing task (left panel) and a chord diagram created from the same data set

This type of visualization is also frequently seen in studies utilizing the classic Visual World paradigm, where multiple objects are presented on the screen while the participant is processing auditory stimuli. There is no ready-for-use Python tool for this type of data visualization, but there are quite a few R packages that could be used, for instance, *VWPre* [10] and *eyetrackingR* [11].

Interest areas are also useful in visualizing the transition of fixations between objects or regions. If the temporal aspect of the transitions is not critical, a chord diagram can be used to visualize the transitions. The left panel of Fig. 8.7 shows the scanpath from a typical experimental task in which the participants viewed a picture depicting a happy bear family flying a kite together. We have five interest areas representing the scene regions occupied by the bear father, mother, sister, brother, and the kite. As is clear from the figure, the scanpath is quite complex, and it is hard to tell if there is a clear pattern in the transitions between the interest areas. In this case a chord diagram will help to reveal the pattern of the transitions (Fig. 8.7, right panel).

To create a chord diagram, you need a matrix that describes the transitions (connections) between the interest areas. With the *chord* module, creating a chord plot requires only one line of code. The resulting chord diagram (see Fig. 8.7, right panel) map is very telling. Most of the transitions were among the bears; there were few transitions from and to the kite.

[10] https://rdrr.io/cran/VWPre

[11] http://www.eyetracking-r.com/

```
#!/usr/bin/env python3
#
# Filename: chord_diagram.py
# Author: Zhiguo Wang
# Date: 4/28/2021
#
# Description:
# A chord diagram to capture the transitions between interest areas

import chord


# The co-occurrence matrix
trans_matrix = [[0, 3, 1, 4, 1],
                [3, 0, 3, 6, 1],
                [1, 3, 0, 9, 1],
                [4, 6, 9, 0, 0],
                [1, 1, 1, 0, 0]]

# Column and row names for the transition matrix
ia_label = ['Brother', 'Mother', 'Father', 'Sister', 'Kite']

# Create a chord diagram and save it to an HTML file
chord.Chord(trans_matrix, ia_label).to_html('transition.html')
```

This final chapter briefly discussed the EyeLink data structure and a few frequently used data visualization techniques. Due to space limitations, we did not delve into the popular data science libraries we used in the example scripts, e.g., NumPy, Matplotlib, and Pandas. If you would like to learn more about these libraries, the *Python Data Science Handbook* by Jake VanderPlas is what I would recommend. The author has generously made this book freely available online, https://jakevdp.github.io/PythonDataScienceHandbook.

The availability of wearable eye trackers has fueled a strong momentum in real-world eye-tracking research. While task design and programming are less of a concern in this type of study, data mining is critical to this type of study. The Pylink library is a task-focused library that helps to build laboratory studies. I would not be surprised to see a data-focused library tailored for real-world eye tracking in the near future.

References

N.C. Anderson, F. Anderson, A. Kingstone, W.F. Bischof, A comparison of scanpath comparison methods. Behav. Res. Methods **47**(4), 1377–1392 (2015). https://doi.org/10.3758/s13428-014-0550-3

T. Blascheck, K. Kurzhals, M. Raschke, M. Burch, D. Weiskopf, T. Ertl, Visualization of eye tracking data: A taxonomy and survey. *Comput. Graphics Forum* ***36***(8), 260–284 (2017). https://doi.org/10.1111/cgf.13079

A. Bojko, Informative or misleading? Heatmaps deconstructed, in *Human-Computer Interaction. New Trends*, ed. by J. A. Jacko, (Springer, Berlin\Heidelberg, 2009), pp. 30–39

D. Bridges, A. Pitiot, M.R. MacAskill, J.W. Peirce, The timing mega-study: comparing a range of experiment generators, both lab-based and online. *PeerJ* ***8***, e9414 (2020). https://doi.org/10.7717/peerj.9414

W. Fu, J. Zhao, Y. Ding, Z. Wang, Dyslexic children are sluggish in disengaging spatial attention. *Dyslexia* **25**(2), 158–172 (2019). https://doi.org/10.1002/dys.1609

Z.M. Griffin, K. Bock, What the eyes say about speaking. Psychol. Sci. **11**(4), 274–279 (2000). https://doi.org/10.1111/1467-9280.00255

R.S. Hessels, D.C. Niehorster, M. Nyström, R. Andersson, I.T.C. Hooge, Is the eye-movement field confused about fixations and saccades? A survey among 124 researchers. *R. Soc. Open Sci.* **5**(8), 180502 (2018). https://doi.org/10.1098/rsos.180502

F. Huettig, J. Rommers, A.S. Meyer, Using the visual world paradigm to study language processing: a review and critical evaluation. *Acta Psychol.* ***137***(2), 151–171 (2011). https://doi.org/10.1016/j.actpsy.2010.11.003

C.H. Lu, R.W. Proctor, The influence of irrelevant location information on performance—A review of the Simon and spatial Stroop effects. *Psychon. Bull. Rev.* ***2***(2), 174–207 (1995)

C.M. MacLeod, The Stroop task: The "gold standard" of attentional measures. *J. Exp. Psychol. Gen.* ***121***(1), 12–14 (1992). https://doi.org/10.1037/0096-3445.121.1.12

G.W. McConkie, K. Rayner, The span of the effective stimulus during a fixation in reading. Percept. Psychophys. **17**(6), 578–586 (1975). https://doi.org/10.3758/BF03203972

J. Merchant, R. Morrissette, J.L. Porterfield, Remote measurement of eye direction allowing subject motion over one cubic foot of space. IEEE Trans. Biomed. Eng. **21**(4), 309–317 (1974). https://doi.org/10.1109/TBME.1974.324318

M.I. Posner, Orienting of attention. *Q. J. Exp. Psychol.* **32**(1), 3–25 (1980)

K. Rayner, The gaze-contingent moving window in reading: development and review. *Vis. Cogn.* ***22***(3–4), 242–258 (2014). https://doi.org/10.1080/13506285.2013.879084

Z. Wang, *Eye-Tracking with Python and Pylink*,
https://doi.org/10.1007/978-3-030-82635-2

D.D. Salvucci, J.H. Goldberg, Identifying fixations and saccades in eye-tracking protocols, in *Proceedings of the symposium on eye tracking research & applications*, pp. 71–78. (2000). https://doi.org/10.1145/355017.355028

SR Research Ltd, *Programming EyeLink® Experiments in Windows (Python)* (SR Research, Mississauga, 2003)

SR Research Ltd, *EyeLink® 1000 Plus User Manual* (SR Research, Mississauga, 2021a)

SR Research Ltd, *EyeLink® Data Viewer User's Manual* (SR Research, Mississauga, 2021b)

M.K. Tanenhaus, M.J. Spivey-Knowlton, K.M. Eberhard, J.C. Sedivy, Integration of visual and linguistic information in spoken language comprehension. *Science* ***268***(5217), 1632–1634 (1995). https://doi.org/10.1126/science.7777863

Index

Z. Wang, *Eye-Tracking with Python and Pylink*,
https://doi.org/10.1007/978-3-030-82635-2

R

MIX
Papier aus verantwortungsvollen Quellen
Paper from responsible sources
FSC® C105338

If you have any concerns about our products,
you can contact us on
ProductSafety@springernature.com

In case Publisher is established outside the EU,
the EU authorized representative is:
Springer Nature Customer Service Center GmbH
Europaplatz 3, 69115 Heidelberg, Germany

Printed by Libri Plureos GmbH
in Hamburg, Germany